CONTENTS

THE TAPIR'S MORNING BATH

PANAMA

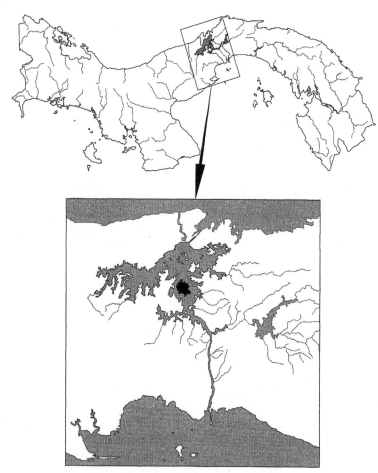

Barro Colorado Island, in the Panama Canal's
Gatun Lake.

1

The Lab in the Jungle

GATUN LAKE, the enormous midsection of the Panama Canal, sprawls for thirty-seven kilometers around peninsulas of land, between fragments of drowned mountains, and over the Continental Divide. Oceangoing vessels slice through the canal and shudder into steel locks that close and open almost silently. The lake's shoreline is wildly irregular, and its waters are as green as the sea.

Impenetrable forest flanks the canal. Toucans screech from low branches, and monkeys leap from tree to tree. Iridescent blue butterflies as large as teacup saucers flit along the shore. Inside the forest, a dark tangle of creeping vines and fringed palms battles to reach the sunlight. Here, where two continents meet and the waters of two vast oceans lap against the lake, lies a teeming cornucopia of life at its competitive extreme, a place like few others on Earth.

From a spot near the middle of Gatun Lake, opposite a deserted village called Frijoles, Barro Colorado Island rises steeply. Its muddy red banks appear jumbled, its interior black. Isolated by the rising waters of the Chagres River, which was dammed in 1910 to form the canal, Barro Colorado had been the highest point on the Loma de Palenquilla ridge. Now the ridge is gone, and Barro Colorado's peninsulas and uplifts sprawl over 1,564 hectares, or six square miles; its summit rises 119 meters above the lake's surface.

From where I stood on the deck of the island's launch as it chugged through the shipping channel, I didn't see Barro Colorado until we were nearly upon it. Then, just before the place where the canal arcs into Bohio Reach, I spotted several red and green channel markers leading into a small cove. A swimming raft floated there. Looking up, I caught a glimpse of tin-roofed dormitories set into the fringed hillside. Emerging from the background of green was a flight of steep concrete steps, which pulled my eye uphill to a graceful veranda and, behind that, to a peaked roof almost lost in the forest's lush canopy.

The low-lying clouds of early morning draped the thickly forested island, giving it the feel of a Chinese landscape painting. Then a small motorboat puttered up to a dock. A woman in camouflage pants tromped across a metal walkway. The lights flickered on in two low-slung buildings. The laboratory in the jungle came to life.

This wasn't my first visit to Barro Colorado. I had traveled to the island nearly ten years before, in 1990, with the much-lauded Harvard biologist Edward O. Wilson. He was there to collect *Pheidole*, the largest genus of ants in the New World; I was there to write about him for a magazine.

A hero to BCI's residents, Wilson was charming and erudite. He'd won two Pulitzer Prizes, for his writing on ants and on human nature, and one Crafoord Prize, the ecologist's equivalent of a Nobel. He'd ushered the subdiscipline of sociobiology into the mainstream, and now, in his sixties, he was lecturing world leaders on the value of conserving biodiversity.

By day, Wilson and I had walked the forest trails. He'd pointed out stingless bees and basilisks, foot-long lizards with craggy fins down their back and tail. He'd explained the intricate relationship between bruchid beetles and a large rodent called an agouti. "Get a load of that," he would say effusively, without a trace of self-consciousness, as he stooped to examine a cryptically colored butterfly.

By night, we had sat around the table in the dining hall, the building with the peaked roof and veranda that overlooked the cove. Over plates of rice and beans a dozen scientists sparred and jousted. They slung statistics and tried to best one another with observations made in the rain forest. "I saw two howler monkeys copulating on Fairchild Trail this morning," a serious-looking plant physiologist said. "I almost stepped on a juvie boa constrictor," a bat researcher countered.

After dinner we drank Atlas beers and the scientists griped about how much money molecular biologists were taking from science budgets, leaving the zoologists, the organismal biologists, the ecologists, with nothing. Names were dropped, tenure decisions criticized. Outside, the jungle thrummed and pulsed; inside, ants streamed over a drop of grape jelly.

Most of the residents were male, with a bias toward entomology. One was studying jumping spiders, another was looking at the flight performance of migratory butterflies, another observed the foraging patterns among leaf-cutter ants. One scientist spent her days examining the teeth of dead anteaters; they offered clues to evolution, she said.

From my first walk with Wilson, the forest had intrigued me. But I found the island residents equally compelling. Like Wilson, they focused on subjects that had seemed, to me, hopelessly arcane. How do frogs produce their mating calls? How much water transpires from a tree? Unlike Wilson, many of the scientists did nothing to hide their ambition. They were often aggressive with one another, or else painfully shy. Many had little social grace. That was fine by me. After all, they lived in a jungle, and their struggle to survive, to use the phrase made popular by Darwin, was tuned to fever pitch.

My first visit to Barro Colorado was brief, but I was there long enough to see that its residents lived and breathed science through their every waking hour. Their language was data, their currency was scholarly publications, their religion was the creative forces of nature itself. I didn't understand a lot of what was going on, but the work seemed important to me, and noble. At the time, the word "biodiversity" was just beginning to enter the common parlance. Rain forests were going up in smoke, and disappearing with them were storehouses of knowledge and potential new drugs, foods, fuels, and fibers. Scientists like Wilson were preaching the gospel of conservation: every piece of the natural world, from microbes to pandas, matters. Caught up in the excitement of this place, I trusted that scientists like these would reveal, someday, exactly how.

When I got home from Panama, images of the rain forest stayed with me, as did the patter of the postdoctoral students in the dining hall and the roar of the insects outside my cabin door. Years passed. Worldwide, natural areas continued to deteriorate. What was the role of scientists now? At a time when so much was going wrong with the environment, fewer people were being trained to *know* the environment. There were fewer biologists who understood the relationships among whole, living organisms or recognized individual species. Science seemed ever more focused on molecular studies, on parsing genomes and analyzing the expression of proteins. Eventually, I wondered, were we going to lose touch with the world around us by being so fascinated by the world inside us?

And yet in this world made smaller and narrower by technology, researchers were still coming to BCI to make broad studies without thought of profits or patents. They were studying evolution in a forest, not in a test tube or a computer. It may sound hokey, but there were still scientists on BCI who studied nature for the pure joy of it. Their exuberance piqued my curiosity. And so, three months before my own wedding, I said goodbye to my fiancé and boarded a plane for Panama.

The island awoke at dawn with the desperate-sounding screams of a thousand howler monkeys, bellowing their territorial yawp. The toucans and parakeets and kiskadees were well up by now, ascreech and atwitter. Bands of coatimundi, their ringed tails held aloft, snuffled through the leaf litter of the lab clearing. In rubber sandals that slapped against the concrete walkways, the scientists slouched downhill toward breakfast.

I'd been here a day already, and I was eager to get into the forest. Alone, I climbed the concrete steps that led away from the lab clearing. Within the forest, the morning racket gradually settled to a low hum of birds, insects, and frogs. An anonymous creature obscured by the tangled understory let loose a sound like broken ceramics in a bag. A chicken-sized bird produced an eerie wail — like a finger circling the rim of a crystal glass.

Moving along the trail, I stepped over ants carrying bright green leaf fragments. I stared at a pattern of brown-dappled light that beamed across the forest floor. It slithered away when I approached — a five-foot boa constrictor with no taste for confrontation.

The forest was greenly dim. The air smelled of dampness, of earth, of mammals. A branch snapped above my head, but no pieces made it through the snarl of vines, saplings, and shrubs to reach the ground. I came upon a fig tree, its bark smooth and its trunk skirted with enormous buttresses. The renowned scientific traveler Henry Walter Bates, working his way through Amazonia in the middle of the nineteenth century, compared these buttress chambers to stalls in a stable, some of them large enough to hold a dozen people.

Woody vines called lianas looped over the forest floor like cursive writing run amok. Grasping neighbor trees with their tendrils, thorns, hooks, arboreal roots, and leader shoots, they hoisted themselves into the canopy, where they lounged over the treetops and sprawled for hundreds of meters. Their stems, meanwhile, grew as thick as many a temperate tree. Examining the fresh tips of one vine, I thought that if only I could sit still for two hours I'd certainly see them grow.

But it was too hot to stay in one place, and soon I moved on,

BARRO COLORADO ISLAND

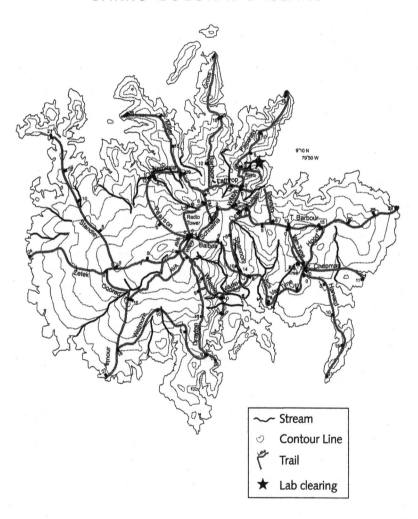

~	Stream
◌	Contour Line
⌐	Trail
★	Lab clearing

turning from Wheeler Trail onto Barbour-Lathrop. A twenty-centimeter seedpod covered with thousands of tiny spikes caught my eye. A spider disguised as white rootlets lay flat against a tree trunk. A blue morpho with a fifteen-centimeter wingspan flopped by in the soggy air, headed downhill toward a sunlit creek. With its arresting coloration and outsize proportions, this butterfly seemed the quintessential symbol of biological weirdness spawned by the hothouse climate. Here things got large, even unseemly: flower petals the size of cake plates, beetles like grenades, leaves as long as coffee tables.

The morpho alit on a tree trunk, folded its wings, and instantly disappeared, its underwing coloration a perfect crypsis against the mottled bark. This was nature, I thought, at the height of her creative powers.

Charles Darwin knew intuitively that tropical forests were places of tremendous intricacy and energy. He and his cohort of scientific naturalists were awed by the beauty of the Neotropics, where they collected tens of thousands of species new to science. But they couldn't have guessed at the complete contents of the rain forest, and they had no idea of its value to humankind. Even now, more than a century later, the mechanisms of the rain forest still baffle, and impress, scientific thinkers.

Some of the best of them have worked on Barro Colorado Island. The laboratory on the island's northeastern shore has operated continuously since 1923, its backyard the most-studied tropical rain forest in the world. Barro Colorado is both a monument of nature and, perhaps more tellingly, a monument *to* nature — off-limits to the general public, virtually stateless. Sitting between two continents, it is populated by field researchers from around the world and administered by the Smithsonian Institution, which acts as a diplomatic mission to science.

The station was the brainchild of James Zetek, a U.S. Department of Agriculture entomologist who'd been working on mosquito control in the Canal Zone since 1909. Zetek, a Czech from Nebraska, had noted the ongoing destruction of the local watershed: land that had been forested was being logged and farmed for the simple reason that it was now, via the canal's labyrinthine shoreline, reachable.

Zetek took every opportunity to speak with scientists who passed through the Zone about setting aside land as a "natural park," but it wasn't until March 1923 that he got lucky. He met up with William Morton Wheeler, a professor of economic entomology at Harvard's Bussey Institute for Research in Applied Biology, and took him by train to the tiny lakeside town of Frijoles, from which point a boatman ferried the men to Barro Colorado.

Wheeler spent just an hour on the island, but in a clearing of less than one acre he collected nineteen species of ants. Zetek took ten species of termites, and each of them took a dozen species of myrmecophiles and termitophiles. "Two new genera, one a beetle, very remarkable!" Wheeler would later write.

Zetek made a similar pitch to Thomas Barbour, the associate curator of reptiles and amphibians at Harvard's Museum of Comparative Zoology, who was also conducting research in Panama that month. Together they decided that BCI, the only large piece of relatively undisturbed virgin forest left in the Canal Zone, would make an ideal place to conduct biological research.

Seeking protection from settlers, hunting, and other human interference, Zetek presented his idea to the Canal Zone's governor, Jay Johnson Morrow, who received it warmly. With an alacrity unheard of in the modern conservation era, Morrow proclaimed the island a nature reserve on April 17, 1923. From this day on, settlers would decamp; any hunters who were found trespassing would be considered poachers.

Unfortunately, Morrow had no funding for the station, and neither did the U.S. government. The men who dreamed of a field station would have to build it themselves. Barbour had recently made a killing on the stock market and was willing to help out. David Fairchild, the U.S. Department of Agriculture's chief "plant explorer," gave his own money (his wife was the daughter of Alexander Graham Bell) and raised even more from his socialite friends Allison Armour and Barbour Lathrop. Their money put up buildings and a track and engine to hoist supplies, by cart, up the 196 concrete steps between the lake and the clearing.

It was here that the laboratory rose. Facing northeast, the wood-frame building afforded excellent views out over the lake, toward the jumbled green hills in the middle distance and, on a clear day, on to the low spine of the Cordillera, the backbone of Panama's uplift.

On my second trip to the island I was happy to see the old lab still standing, though substantially reconfigured; it looked trim and neatly painted. I'd eaten in the downstairs dining room ten years before, but the building was now used as a visitor's center and a party hall. Since a wave of renovations in the early 1990s, everyone ate in a new building farther downslope. The scientists worked in air-conditioned labs near the lakeshore and slept in relatively insect-free dorms built of poured concrete.

Beyond the lab clearing, though, everything seemed the same — as ten years earlier, or a hundred. The island was still thickly forested, and there were still no roads or villages. Evidence of modernity was scant. History looped all around me in the fifty-nine-kilometer trail system.

Turning off Barbour-Lathrop, I was walking through time, along trails that formed an epic poem of the research station's, and of tropical biology's, history. Zetek, Wheeler, Barbour, Armour, Fairchild, Gross: these were the names of the men who built the station, who made its reputation. The dusty specimen jars of BCI's old herbarium and the abandoned monkey cages rusting along Allee Creek punctuated this history, reminders of the station's evolution from the days of pure plant-and-animal description through the advent of naturalistic studies of creatures in their native habitats.

Downhill, in the modern laboratories, gas analyzers and slides holding snippets of DNA pointed to the future. These new methods, in concert with the old, were shining light on the inner workings of the tropical forest — the most elaborate and complicated natural system on the planet, home to two-thirds of the approximately 4.5 million species alive on Earth today.

BCI is dissected by radial streams that flow, in the rainy season, down its steep ravines. The ubiquitous ridges and gullies can make for rough going. There isn't a sizeable section of level trail on the entire island, except for an area near its plateau, where I turned onto a trail named for Allison V. Armour.

A philanthropist, Armour liked being around scientists enough to shuttle them to research sites around the world on his yacht, the *Utowana*. He contributed tens of thousands of dollars to the field station during its early years and donated a launch, called the *AVA*, to the island.

I knew I was the first one down Armour that day because I had to break trail — bust up the thick, sticky webs of *Nephila*, a psychedelic giant. *Nephila*'s body is about two and a half centimeters long, with iridescent green and red markings. Her eight legs are long, lacquer shiny, and delicately pointed. After blundering into several webs and wondering if the builder — who isn't actually dangerous — was now inside my T-shirt, I plucked a stick and walked like a conductor wielding a baton, with particular emphasis on the downbeat.

I began to notice the local tree bark. I had no hope of identifying anything. The leaves were usually too far up to see, or too entwined with other branches for me to tell which belonged to which. A mere fifty hectares at the top of the island contain more than 300 tree species, more than are found in all of North America above Mexico. The island itself has 1,369 species of vascular plants — more than in all of Europe.

Most of BCI's trees have smooth bark, or bark mottled with shades of gray or brown. Some trees are ponderously columnar. Others, wrote Frank Chapman, an ornithologist who spent twelve consecutive dry seasons on the island, "suggest stripped athletes, with muscles and sinews swelling beneath their thin skins."

Although the northeastern half of the island had been widely cut in the late nineteenth century, the "back side," where I was now, contained trees more than a hundred years old on land that had been little disturbed for centuries. Many of these trees have large buttresses, which give them circumferences of more than fifty feet.

How old are these grand specimens? It is difficult to say. Without a distinct growing season, tropical trees lack growth rings. In general, they have shorter life spans than do their temperate cousins. They fall victim to pathogens and herbivores, they have shallow roots, and heavy winds and clinging vines often topple them.

For all that was known of Barro Colorado Island, for all the measuring, observing, comparing, and thinking over seventy-five years, the forest did not give up its mysteries easily, which was why it attracted a steady stream of investigators. While academics in the United States debated whether science had reached an end, whether all our great discoveries were behind us, researchers on Barro Colorado didn't seem to have heard the news. The island, home of the world's longest-running seminar on tropical biology, had never been busier.

In time I would learn to identify a fair number of plants, but for now they confused me. Like most nonscientist visitors, I was far more interested in BCI's animals. Isolated for centuries by rough terrain, thick jungle, and diseases that kept man at bay, then protected by Gatun Lake and legal decree, Barro Colorado nurtured some of the most diverse animal life of any tropical nature reserve.

In the months to come I would regularly spot squirrels, rats, peccaries, agoutis, coatimundi, deer, sloths, monkeys, tayras, anteaters, bats, iguanas, geckos, basilisks, crocodiles, and dozens and dozens of gaudy birds. If I had applied myself, I might have seen a tapir or an ocelot.

Early scientific visitors to the tropics had the thrill of naming thousands of creatures. Henry Walter Bates, for example, came out of the Amazon with 14,712 species, almost two-thirds of which were unknown to science. As we enter another millennium, the contents of the Neotropics, even a place as well traveled as BCI, are still being described.

I walked on, past a small metal sign marked AVA 4. It meant I was four hundred meters from the trail's beginning. All island trails are marked this way, every hundred meters, which makes it easy to find things and hard to get lost. Each of the island's fig trees is also marked and mapped, and atop the island's plateau, every stem larger than one centimeter in diameter at breast height (1.3 meters from the ground) wears an identifying number. Had everything on this island been examined and mapped? I wondered.

One day a few months on, I would find myself in an area of BCI that had no tags or flagging, no numbers or stakes or signage — a place where the hand of Science had yet to leave a mark. The area felt remote, silent. There were no trails, and I had to navigate by compass. No part of the island had ever felt so wild to me.

As a child, I took pleasure in imagining that I was the first person ever to walk the woods beyond my suburban house. As an adult, far from BCI's lab clearing, I easily conjured up that feeling once again. But the feeling didn't last. After twenty minutes of bushwhacking I spotted a plastic flag in the ground. My heart sank. Then I spotted a tree with a metal tag, then another and another.

With a pang of disappointment I realized that all here had been inventoried, that these acres were as studied as any others on the island. Someone within a few kilometers of me could name (and probably had) every plant in sight. Someone else could name all the vertebrates. As for the invertebrates, well, those descriptions were on their way. Entomologists had about a million species sussed — most of which occurred in the Tropics — with several million more just waiting to be described. That science knew so much about tropical organisms was a good thing, to be sure. But I liked a little mystery in nature, a little unruliness, all the same. This idea — that there was something left to discover out there, whether the names of new organisms or the processes that allowed so many to coexist — would come to thrill me as the months went by.

Now, near AVA 5, a small brown creature, sleek of fur and delicate of foot, scuttled across the path. It was an agouti, which looks like a rabbit crossed with a deer but is actually a rodent, a fact I was reminded of as it eyed me blandly and perched on its hind legs, squirrel-style, to gnaw on a large seed.

The unique fauna of Central America confused the earliest non-native explorers. Columbus reported seeing lions, stags, and gazelles here — creatures that sprang from an Old World mindset. John Lloyd Stephens, an American booster of a sea-level canal, saw alligators and heard both wolves and mountain cats, creatures of North America, on his 1840 journey through Nicaragua.

Hoping to find a dead animal, something I could examine close up, I walked on. Lizards scuttled through the dry leaf litter. Purple blossoms wafted onto the forest floor. Dull-colored understory birds pecked at insects. What kinds of lizards, flowers, birds? I had no idea. Alexander von Humboldt, when he first visited the Venezuelan rain forest in 1799, expressed his frustration at recognizing so little of this landscape: "We rush around like the demented; in the first three days we are unable to classify anything; we pick up one object, to throw it away for the next."

Darwin and Bates, as well as Richard Spruce and Alfred

Russel Wallace — nineteenth-century British naturalists inspired by von Humboldt's accounts of the Neotropics — felt similarly overwhelmed by the forest's abundance. Its beauty and complexity appealed to their aesthetic senses, but as scientists they yearned to *classify*. Compared to the temperate forests, the tropics had so much more of everything: more butterflies, more beetles, more yellow birds. Why, these naturalists asked, were there so many species? As Darwin put it, "What explains the riot?"

Rain forests, as everyone knows, are all about competition — tightly packed species scrambling for food and sex and light and turf. Organisms employ ingenious methods of outsmarting one another — crypsis and camouflage, chemical deterrents and chemical attractants. Parasites drive some bats to live as singles and the seeds of trees to grow wings, the better to escape the crowds that invite attack.

Yet for all this competition, different species live side by side, many depend on others for their very existence, and most owe their competitive edge to the constant pressure of their rivals. How did such systems evolve? Are there general principles that explain the maintenance of the forest's diversity? How do the forces of cooperation, or mutualism, measure up to those of competition? Increasingly, it's cooperation that scientists emphasize in their evolutionary studies.

Questions about the origin of species and the persistence of species stand at the heart of evolutionary and ecological studies today. The answers have value as basic research, and they are critical for conservation planning as well. Some trees, it turns out, are pivotal in maintaining a community of plants and animals, and their loss could cause a chain reaction of other losses.

Darwin's question, about the relative abundance of the tropics as compared to the temperate zone — a phenomenon known as the latitudinal diversity gradient — remains to this day one of biology's thorniest questions. Scientific thinkers over the decades have posited numerous theories, both complex and simple. Most of these existing theories are untestable, some contradict others, and not one is accepted as dogma. The gap in understanding left room for a young scientist I'd soon meet to puzzle out a new theory and introduce it in the marketplace of scientific ideas.

I meandered slowly through the wonderland of the island's western slope, the green light shifting and fading to gray as the afternoon wore on. The intermittent breeze carried a cornucopia of scents: wintergreen, sugar, skunk, monkey house, maple syrup, air freshener, sweat. Underfoot, twelve centimeters of leaf litter crackled. By the middle of the rainy season it would all be gone, the earth bare.

I hiked on, though paranoia dogged me. Did these leaves swarm with chiggers, and were they working their way into my socks? I was tempted to stop and investigate, but this was the first hike of my field season, and I was trying to set a standard of physical stoicism. At last I reached Armour's end, at the lake. But this was no beauty spot. The shoreline was boggy; the lake was picketed by white, eighty-year-old treetops, dead but still standing. It looked like a cemetery.

No breeze reached this desolate place, which evoked in me a sort of pre-canal jungle horror. It was just this situation that David McCullough described in *The Path Between the Seas*, his chronicle of the canal's creation: "All varieties of tropical fever and miasma were caused by 'noxious vapors' released from the putrid vegetation and rank soil of the jungle. Any excessive disturbance of such ground, therefore, naturally meant the spread of disease in epidemic proportions."

Contemplating the potential for miasma, I sat down on a handy rotted log, then almost immediately leapt up. The log was crawling with seed ticks the size of a pencil point. I spent twenty minutes plucking several hundred of them off my legs with a scrap of duct tape I found in my backpack. Then another ten minutes wrapping my duct tape in duct tape, so I could transport the tick ball back to the lab. What a magical place is a rain forest, I thought to myself.

The tropical rain forest, like the polar regions or the planet's great deserts, is largely a metaphorical landscape. In fiction, in art, and in poetry, it represents fecundity, sensuality, and wonderment. A hundred years ago, at the height of the tropics' scientific exploration, the rain forest's endless growing season and its diversity of plant and animal life implied to those coming out of the comparatively sterile north — with its limited roster of plants and animals, its orderly gardens, and its bleak weather — both a primeval loucheness and proof of a magnificent life force. Darwin called Brazil's rain forest a luxuriant hothouse, and one suspects he meant it in more ways than one.

The South American forests, undefaced by the hand of man, were "sublime," Darwin said, a temple "filled with the varied productions of the God of Nature: — no one can stand in these solitudes unmoved, and not feel that there is more in man than the mere breath of his body." Of his meanderings in the forest, he wrote in his journal, "I never experienced such intense delight."

For travelers less scientific, the Paleo- and Neotropics represented a place to get lost, a heart of darkness within which the trappings of civilization could be forgotten. The setting was appropriate — dimly lit, uncatalogued, unmapped, and populated by lightly clad primitives. Bizarre life forms advertised their sexual fitness with gaudy colors and displays. Here, everything grew more rapidly: parasites lived on parasites, flowers lingered for months at a time. Decay, too, was quicker: mold, insects, and fungi pounced on the heels of death. Together, they made the renewal of life all the more swift.

Many European naturalists visited Central America in the seventeenth and eighteenth centuries, but none had the intellectual background to address scientific issues: Why were New and Old World organisms different, or similar? Even by the mid-nineteenth century, only a handful of people with a Western scientific perspective had entered the tropics. There was no uniform theory explaining how life on earth had evolved; the Amazon's biota was entirely unclassified. For men like Darwin, Bates, and Wallace (who visited South America but didn't make it to Central America), the field was wide open.

Their experiences in the tropics would become essential to these British naturalists' later formulations of evolutionary theory. Upon noting the wide dissemination of plant and animal species along an Amazon tributary, for example, Bates began to wonder how their seeds and eggs got around: "Unless it can be shown that these [species] may have migrated or been accidentally transported from one point to the other, we shall have to come to the strange conclusion that the same species had been created in two separate districts." This idea would later come to be known as convergent evolution, which explains how unrelated organisms independently develop similar features in response to similar environmental challenges. Such traits include how to locomote, how to conserve water in a hot environment, or how to store energy.

Working in the Malay Archipelago, Wallace wrote in 1848: "There must be some other principle regulating the infinitely varied forms of animal life. It must strike every one, that the numbers of birds and insects of different groups, having scarcely any resemblance to each other, which yet feed on the same food and inhabit the same localities, cannot have been so differently constructed and adorned for that purpose alone." In short, Wallace was asking why competition doesn't exclude all but the fittest creature from its niche. He was asking the eternal question: How do similar species coexist?

The nineteenth century both demonized the rain forest as a place of dank mystery, bad morals, and disease, and romanticized it as a garden of abundance and innocence. More recently, the rain forest has come to symbolize man's destruction of the planet. As goes the rain forest, according to this dictum, so go we.

Developing nations use their disappearing rain forests to gain geopolitical leverage; tribes indigenous to the rain forest play on First World ecoguilt to rally support for their threatened way of life. Environmental groups raise money off images of rain forest destruction; ecotourism operators lure well-heeled vacationers with images of its beauty and abundance. Here on BCI, doctoral candidates ignore the aesthetics and the politics of the forest in order to make of it a proving ground, a place to enact the research that leads to Ph.D.s, then jobs.

For those not living in a remote jungle, however, politics is hard to ignore. Every day comes word or image of a natural landscape in decline. We are living, as David Quammen has written, in one of the great soul-searching moments of human history. Never before has our fate, linked inextricably with the health of the planet, rested so surely in our own hands. Indeed, it is now the actions of human beings themselves, and not solely the blind mechanisms of natural selection and genetic drift, that are altering the terms of survival for every species on the planet.

The current extinction crisis — in which an estimated one-third of one percent of all bird and mammal species go extinct each year — is on course to rival that of the Big Five, of the Ordovician, the Devonian, the Permian, the Triassic, and the Cretaceous. How we respond to species extinction, and to massive habitat loss, global warming, and unchecked air and water pollution, will likely define the new century.

It was an exciting time to be among scientists, among field biologists working at ground zero. I had come to Panama to see what that work looked like, and how studies so seemingly arcane could possibly shape decisions yet to be made. I had come here, also, because nature had enthralled and confounded me from an early age and because here were people who could decipher it.

As I walked back to the lab from Armour Trail with tick ball in hand, my enthusiasm for this place remained undimmed. I was going to see an intact rain forest, through the eyes of experts, before all the rain forests were gone.

2

Pieces of the Frame

I UNPACKED SLOWLY over the first few days, folding my sparse wardrobe into the dry closet. This was a regular wooden closet that contained a plug-in heating rod. It worked reasonably well in the dry season, but when the rainy season set in, nothing would stop the mold from colonizing my shoes, my wallet, anything made of leather. Heavy cotton clothing remained in a perpetual state of dampness.

When things got busy at the field station, residents doubled up in the dorms. For now, though, I was alone, and my room was wonderfully spartan. It contained two lamps, two beds, two desks, and two chairs. A ceiling fan creaked overhead, and moth larvae inched over the concrete walls, which were bare except for a poster of an emperor penguin.

As I settled in, I reviewed my goals. The plan was to live at BCI, off and on, for a year, reporting on the work of the scientists and on the daily life of the field station. I announced my intentions to a skeptical audience at an island meeting, then decided to lie low for a while, to get the feel of the place and, I hoped, to earn the residents' confidence.

I came up with a modus operandi: I'd offer myself as a field assistant and give free help to anyone who asked. Fieldwork was often grueling. I was happy to pitch in, and the work would help me learn about being a scientist, day in and day out, and about how a forest was studied.

Just about everything was new to me on BCI, and getting familiar with the lab clearing, the island's human hub, would take a little time. Oris Acevedo, the island's scientific coordinator, gave me an introductory tour. A native of Panama, with short, dark hair and a laid-back sense of authority, Oris had worked on BCI since 1992. She had a master's in forest management and was fluent in both English and German. The latter was a big plus, since Germans were ubiquitous in the plant and bat departments here.

BCI's six modern dorms were arrayed on the hillside not far from the shoreline. Seen from the canal, with their turquoise roofs, they looked vaguely Mediterranean. Seen close up, they looked Soviet. High above the lab clearing stood three older dorms. Altogether, there was room for thirty-five residents.

"You know the dining hall," Oris said, gesturing toward the upstairs level of the central concrete building. Meals appeared, buffet style, at six-thirty, noon, and six-thirty. The dining hall also contained the mailroom, which in turn housed two telephones. In general, if the lines weren't busy, the telephones were broken.

"Downstairs is the conference room," Oris said. We poked our heads into the air-conditioned auditorium. Residents presented their "Bambis" here — casual Wednesday-night seminars on their current research. In contrast to the weekly salons that postdocs used to hold, discussing the latest journal articles on tropical biology (postdocs have their Ph.D.s but continue to gain field experience, on fellowship, before seeking university jobs), these gatherings were supposed to be low-key and as unthreatening as an animated fawn. Recently, though, the rigor quotient of Bambis had begun to creep up, and residents now presented massive amounts of complicated data that often overwhelmed the island's younger residents. Not to mention me.

Oris and I walked downhill toward the two lab buildings, called north and south, which were connected on the second floor by a walkway. Constructed of concrete and well equipped, they wouldn't have looked out of place on a modern university campus. The more permanent labs had names on their doors — Stingless Bee Lab, Water Lab, Fruit and Seed Lab, and Psychotria Lab, after the island's most common shrub.

We saw the darkroom, the balance room, the chemistry lab, the herbarium. This last room would become a favorite of mine. In addition to shelves of reference books, it contained sheaves of plant voucher specimens and tray after tray of pinned butterflies, which were beautiful to look at and enormously helpful when I was trying to identify creatures I'd seen in the forest.

Water coolers, emergency eyewash stations, and Gary Larson cartoons — the usual stuff of American science buildings — filled the hallways. Bulletin boards advertised job opportunities around the world. Maps were popular, and bar graphs of daily and monthly levels of solar radiation, relative humidity, soil moisture, and cumulative precipitation fetishistically festooned the walls. In the south lab building hung a poster that always made me smile. Above drawings of various mushrooms, some poisonous, it read, "Have fungi, but be careful."

On the first floor of the north building, past a pile of rubber boots and a shipping crate shaped like a coffin, Oris swung open the door to the last lab on the hallway. "You can work in here," she said.

It was a big room, dominated by soapstone lab benches covered with equipment. I'd have two sinks at my disposal, plus a fume hood, gas mixer, soil borer, quantum radiometer, chlorophyll fluorometer, infrared analyzer, constant-temperature circulator, field pump, tank of liquid nitrogen, and a hundred pounds of twelve-volt batteries and thermocouples. I didn't know how half these things worked, but I liked the professional ambience they provided.

On the wall over the desk, some sentimental northern soul had glued an array of deciduous leaves: sugar maple, white oak, sweet gum, beech, and holly. The leaves made me feel at home. I placed my computer on the desk and plugged it in. My lab was up and running.

At the tour's end, Oris pointed out the island's lounge, a large room at the end of the south lab building. It was late afternoon and two residents were sprawled on ratty furniture reading back issues of *Nature*. There was a TV, which didn't get any stations, and a VCR. The kitchen was modestly equipped. Food-encrusted plates filled the sink, and grease coated the stove.

My tour was over, but for some reason Oris hadn't shown me BCI's bar, on the balcony outside the lounge. I stumbled on it by myself the next morning and found a few plastic chairs and a church key tied to the railing. Bottle caps on the floor and cigarette butts in the ashtray indicated that the place was popular, but I wouldn't find out exactly how popular until a week or so had passed.

On my first trip to BCI I'd felt perfectly comfortable drinking beer with unknown scientists, but I'd had E. O. Wilson for my chaperone. Now, having not yet made any friends, I had no pretext for visiting the bar, and the prospect of facing a crowd by myself intimidated me (just as I, a notebook-toting outsider, intimidated the crowd). Fortunately, there was soon another social locus, a place far more exclusive, for me to do my drinking: the apartment of fifty-eight-year-old Egbert Leigh, the island's official scientist in residence.

Theoretically, it was easy to meet people at mealtimes, when they were relaxed and available for conversation. I'd been watching a redhead named Christina Campbell for two days. She sat at the middle of the long dining table, always facing the trolleys of food. Most people preferred the view of the lake, but Chrissy wanted to see who came in when and with whom. She was surrounded by a gaggle of field assistants, monkey researchers from Berkeley. I wanted to get friendly enough with her to ask about the island's spider monkeys, her study subject, but I didn't get far.

"Where are you from?" I asked.

"New Zealand," she said, and turned pointedly away. I was more amused by her rudeness than hurt, since I'd done nothing so far to provoke it.

Bret Weinstein rarely made it to breakfast or lunch, and he came to dinner late, after most people had finished eating. He worked at night, trying to figure out why certain fruit-eating bats constructed tents from leaves. Bret was of medium build, with longish wavy hair that he kept out of his eyes with a bandanna. He had a reputation for being smart and disorganized. People were always asking him questions — about animals, the field station, people. He'd give ridiculous answers that sounded plausible, gauge his interrogator's reaction, then go straight. I liked him right away, but it would take weeks for us to feel comfortable enough to work together in the forest.

Hubert Herz was a gimlet-eyed German who studied leaf-cutter ants, the creatures I'd seen waving parasols of green along the trail. He wanted to know how ants affected trees, whether ants changed the rate at which trees exchanged gases and grew new leaves. I was drawn to Hubi's boyish enthusiasm for everything around him. He'd plop down at the lunch table, sweat-soaked and muddy, then tell everyone about the treefall he'd witnessed or the iguana that had leapt into the lake.

Egbert Leigh, who would soon become my new drinking companion, arrived at meals promptly, sat facing the great outdoors, and inquired rather formally after the progress of some, but not all, of the researchers. A thin man with a belly, white-bearded but brown-haired, he was the island's elder statesman and most esteemed scholar. He avoided small talk and toiled in his book-lined apartment next to the laundry room. His subject was mutualisms, the intricate symbiotic relationships between living organisms. A theoretician, he ventured into the forest but rarely. About half the chairs at the scientists' table were occupied by field assistants, nearly a dozen of them. A few older plant physiologists, mellowed and sage, held down the end of the table, and a clique of Germans floated freely but as one. Like any captive group lacking control over its diet — summer campers, stranded explorers — the residents obsessed over what they ate and complained fulsomely. "Why can't we get any yogurt here?" they'd ask. "What is this form of animal protein?" But food gripes were largely a put-on, because no one here was seriously deprived. Most field biologists aren't served three meals a day, or even one, which those who'd roughed it at other field stations were quick to point out.

Having someone like me record the station's comforts made not a few residents squirm. They had peers in squalid field camps around the world who lived without hot showers, air-conditioned labs, or e-mail. They didn't want me, an outsider, fueling the rumor that circulated around American biology departments, that BCI is a Club Med among research stations.

Despite its comforts, though, BCI is a research station in the classic sense: an outpost that allows individuals to function in the field as scientists, not expeditionaries. Dozens of such places dot the globe, from one-room shacks in the mountains of North America to tent camps in the African veldt, from high-tech laboratories in Indonesia to rudimentary setups in the rain forests of South America. The best-equipped field stations leave scientists relatively unfettered, providing meals, a water source, labs, and, usually, a steady source of electricity.

A century and a half ago, naturalists studying the flora and fauna of the Amazon carried ammunition, guns, game bags, nets, boards and materials for skinning and preserving animals, a botanical press and papers, drying cages for insects and birds, various chests and storage boxes, natural history books — and a lot of money, to pay servants to haul this impedimenta around. The pioneering Richard Spruce, collecting specimens along the waterways of Venezuela, found the region most unfriendly to travelers and thus to science. Indians threatened to rob and kill him. As he traveled he slept in hammocks, boats, and huts infested with rats, scorpions, cockroaches, and vampire bats. Food was scarce along the Uaupes River. His time, he said, was consumed "in procuring materials for a miserable existence."

In the modern era, many a tent camp erected by an energetic individual has evolved into a bona fide field station as word spreads among scientists. Over the years BCI has become a crossroads, a place where researchers come across other scientists they wouldn't ordinarily meet or interact with — plant physiologists and geologists, anthropologists and biogeochemists, palynologists and ecologists, molecular biologists and archaeologists, and even, while I was on the island, a creationist.

Thanks to those who've gone before them, today's researchers have at their fingertips data on thousands of different life forms gleaned over seventy years. Some data go back even farther. The French began measuring Panamanian rainfall and other environmental factors for potential canal operations in the 1870s. They quit the isthmus before the turn of the century, but the Americans' Panama Canal Commission soon picked up the slack.

Before they even set foot on BCI, researchers know what the island's soil is like, how many of their study subjects live here, and what they prefer to eat. Food sources have been quantified, trees mapped. Building on this foundation of baseline data, investigators can design subtler and more sophisticated analytical studies. Equipment for collection and analysis is in place, the island's administrators arrange for any needed permits, and supplies are just a day-trip away. A researcher can arrive on Monday and start logging data on Tuesday.

While I myself settled in at BCI, I collected stories of researchers who actually *had* lived rough in pursuit of field data. During his field period in the Amazon, one herpetologist managed to avoid Chagas disease — the potentially fatal illness transmitted by the assassin beetles that infested his shack — but spent three months too sick to work, from eating food cooked by natives. I met a female scientist harassed by local men, a researcher deliberately led astray by locals in the search for her study organism, and others who spent months, even years, trying to obtain permits to work in a foreign country, to collect and export specimens.

Where a scientist works has much to do with the type of work he or she can do. In rudimentary field sites, the work tends to be more descriptive than analytical: if you can't freeze tissue samples, you can't work with DNA.

David Marsh, studying tungara frogs on BCI, had spent three months in a tent in Ecuador, where he studied reptiles. "But I didn't get nearly the amount of work done that I would have on BCI," he told me. "For one thing, it took me four hours to wash my clothes, which were always filthy, and they never dried." Every couple of weeks he had to leave his camp to buy food in town, which involved several kilometers of hiking and then a long ride in a truck. One night Marsh left his tent for a pee, and bullets whistled by his head. "The local peasants were shooting at me," he said. "They thought I was poaching their charcoal.

"So yes," he concluded, "some people do think of BCI as a country club. But a fully operational place contributes to the quality and the quantity of work produced."

In the upper lab clearing, BCI's past lingered on. The old dining hall sat prowlike on the hillside, and the low-roofed Haskins, a former lab turned storage area, lurked damply in its shadow, just barely free of the entangling forest. Off the concrete path sat Smith, Kodak House, and ZMA, wooden dorms now occupied by that tiny subset of residents who were both nostalgic for the past and hardy enough to mount the hundred-odd steps several times a day. Roach-infested and rough, these buildings were metaphors for BCI's pioneer days.

In the old herbarium, hard by Haskins, yellowed index cards filled a cabinet. Taped to them by James Zetek's own hand were arachnid specimens seventy years old. In jars of formaldehyde swished snake eggs, a sloth stomach, a white crustacean, two thirteen-centimeter-long monkey babies, a mouse. One jar was jammed tight with bats, its label long ago decayed.

Guided by their own enthusiasms as much as by the mandates of science, BCI's earliest visitors collected and classified, laying the foundation for what would become one of the deepest troves of environmental baseline data in the tropics. These scientific papers, which filled more than a dozen bound volumes in the island's lounge, told the stories of BCI's plants, animals, fungi, rocks, and the relationships between them. The early papers described, for example, what a troop of monkeys ate, or how an ant colony established a new nest. In later years the papers would acquire quantitative rigor: how many calories the monkey diet contained, the chemical formulas of the signals emitted by these ants.

The data gleaned on BCI over the years form a continuum, with each new piece of information joining the millions of other pieces already in hand. Seventy-five years ago investigators worked independently, as they do today, but all have shared the ultimate goal of understanding this most intricate of the world's ecosystems.

A jigsaw puzzle makes an apt metaphor for the parsing of this system. Each scientist potentially holds a puzzle piece, though he or she doesn't necessarily know how it fits into the whole or even what pieces other researchers hold. Some puzzle pieces get set aside, at the edge of the work table, until investigators figure out where they fit. Some puzzle pieces are universal and can be used over and over again — general principles about photosynthesis, for example, or reciprocal altruism.

The biggest, most obvious pieces were identified early, and they functioned as the jigsaw puzzle's frame. It was the station's founders who established that the island is sharply seasonal; that its precipitation averages 2,600 millimeters, or 102 inches, a year; and that ninety percent of that falls during the rainy season, which lasts from May until December. The annual temperature averages 27 degrees Celsius, or 81 degrees Fahrenheit, with an average diurnal range of 9 degrees Celsius. Geographers provided another piece of the frame, pinpointing the island's latitude at 9 degrees 10 minutes north and its longitude at 79 degrees 50 minutes west. Geologists determined that BCI's sedimentary formations, seamed with volcanic debris and topped by a basaltic cap, sprawl over 1,564 hectares.

Some baseline data, like species numbers, are periodically updated. At last count, the island had 65 species of terrestrial mammals, 70 of bats, 381 of birds, 58 of reptiles, and 32 of amphibians. Not only are the numbers of species known, but in many cases so is the estimated number of individuals. Over the island's roughly fifteen hundred hectares roam an estimated 2,500 agoutis and between 5,000 and 7,000 sloths.

The field station and its data collectors form a positive feedback loop: the station's comfort and convenience attract researchers who, in turn, churn out more data, which attracts more researchers. Why? Because more information about more organisms adds context — depth and breadth — to their current studies. Bret Weinstein, looking at fruit-eating bats, had only to read BCI's *Flora*, the island's guide to plants, to know the seasonality of the fruit consumed by his bats, the fruit's physiology, and its growth patterns — information that helped him hone his questions.

Having baseline data lets scientists set up comparison studies and pose questions that weren't yet formulated at the time of the data's recording. For example, a plant physiologist investigating a forest's fruit production over the years can look back and correlate her numbers with El Niño events.

Questions large and small benefit from good baseline data. One explanation for the latitudinal diversity gradient, a theory called the stability-time hypothesis, holds that the tropics have been physiologically and climatically stable for so long that many species have had time to evolve here and accumulate. Darwin, among others, subscribed to this idea. But how stable are the tropics, in reality? Without long-term records, scientists can only make extrapolations.

The founders of the field station on BCI discovered early on that running a jungle laboratory was not easy. Vegetation grew quickly. Insects and fungi attacked building materials and provisions. Even cable covered in lead was no match for termites.

The Panamanian weather could be abominable. Moisture wreaked havoc with scientific equipment. "Anything made of iron or steel turned bright orange with rust," McCullough wrote in *The Path Between the Seas*. "Books, shoes, belts, knapsacks, instrument cases, machete scabbards, grew mold overnight. Glued furniture fell apart. Clothes seldom ever dried ... Without laundry facilities, a clean shirt or fresh pair of trousers were luxuries beyond compare."

Against the odds, James Zetek and Thomas Barbour persisted. They bought obsolete building supplies from the Panama Canal Commission and "begged and borrowed," according to Barbour, from the army and the navy. Zetek had been in the Zone for decades, and he knew exactly whom to browbeat to get what he wanted — whether a free water tank or termite-proof buildings.

In 1924, living conditions on BCI were still fairly primitive, but the lab was outfitted with a drying room, a darkroom, scales, microscopes, and a complete set of chemicals and glassware. Workmen brought ice over from Frijoles; lakewater ran from the taps. Zetek laid out trails across the island and hired locals to cut them. Eventually they'd cover twenty-five miles and be named after the station's primary benefactors and distinguished naturalists.

Thomas Barbour arrived that spring, to survey amphibians and reptiles, smoke cigars, socialize, and take the measure of this place he'd helped to build. In some respects he was simply moving south his famous "eaterie" — a lunchtime gathering of garrulous biologists that he hosted in his office at Harvard's Agassiz Museum.

An enormous man, Barbour pulled his pants high over his bulging belly and tucked his cuffs into boots that laced to the knee. He walked the forest trails with notebooks, a machete, and a collecting bag. He slept in a large tent, where he read until two A.M., rising frequently to shine a spotlight on the opossums or night monkeys that pounced on his roof. He'd get up for good at five forty-five, set himself down on the dining hall's screened porch, and watch the white-faced monkeys drink rainwater from the pallid chalices of the balsa-tree blossoms.

Barbour was pleased with what his speculating had wrought. "The building of this laboratory has made it possible for the teacher of biology with a small salary to have the thrill of Wallace, Bates and Spruce when they first set foot in the Amazon jungle," he wrote in his memoir, *Naturalist at Large*. For Barbour, BCI provided all the illusion of exploration but with one great difference: "We have pure drinking water."

Later that summer, William Morton Wheeler arrived to census ants. With him was David Fairchild, interested in the island's fruiting trees, and Fairchild's seventeen-year-old son, Graham. During this six-week idyll, it was Graham's duty to serve "the assembled great men," as he called them, a dawn snack of a banana. While the scientists dressed and ate, they spoke freely. Graham considered these morning sessions most instructive, as they were "devoted largely to discussing the amatory adventures of absent colleagues." The stories, he said, would make your hair curl.

Wheeler, like Barbour, had wide interests. He read Greek,

French, German, and Latin, and he consumed almost a book a day. Such was his enthusiasm for ants that anyone who set foot on the island, from construction workers to students of vertebrates, ended up assisting him. In his characteristically unguarded manner, Fairchild *père* gushed to Wheeler, "You are just the most exciting thing in the whole range of jungle things."

Even before the lab had opened for business, Zetek and his colleague Thomas Snyder were conducting research on the island, studying the protozoan fauna of the termite's intestinal tract, which allowed it to digest cellulose, and trying to determine methods of wood treatment that would withstand termite attack. In a very real sense, Barro Colorado is the field station that termites built. The fact that thirty different species were constantly trying to tear it down attracted from the pest-control industry no small amount of interest, and funding.

In August 1926, the American Wood-Preservers' Association and the U.S. Bureau of Entomology erected Termite House, at the end of Drayton Trail, and Redwood House, at the end of Armour. Built of wood treated with zinc chloride and creosote and outfitted with small oil stoves and beds, these structures were meant to shelter naturalists conducting night studies and also to demonstrate the termite-resistant powers of their timbers.

Apart from the meager subscription fees of universities and the gifts of wealthy patrons — which usually amounted to no more than a few hundred dollars — termite money was the only funding BCI saw for several decades. It was termite money, in fact, that put up the first dormitories, including the moldering ZMA — for "zinc meta arsenite" — still in use today.

Throughout the 1920s, '30s, and '40s, Zetek held down his job with the Department of Agriculture, based in Panama City.

Eventually he came to the island just one or two days a week. His scarceness was just as well, for many of the scientists who lived on the island considered him a petty tyrant. "There was a pack of orders of the day, rules and regulations," recalled Robert K. Enders, a mammalogist from Swarthmore College who began visiting BCI in 1930. "Some of us, I must admit, used to get a little hilarious over his descriptions as to how to take a shower."

Back at Harvard, Thomas Barbour walked in to work every morning and said, "Give us this day our daily Zetek," expecting the usual telegraphed list of complaints and pleas for money. "Everybody that dealt with [Zetek] was frustrated by him," recalled Graham Fairchild. "They loved him dearly, but . . . he was quite cantankerous . . . and if he didn't like you, you were in trouble."

Those who irked Zetek the most were poachers, who continued to take pacas, tapirs, peccaries, ocelots, and jaguars from the island, despite its status as a refuge. Residents who came upon poachers, Zetek wrote in his exhaustive handbook "Information for Scientists," should observe them carefully: "If you are dead still, where he cannot see you, and you have a strong stick or what not, let him pass you and then you give him a good crack across his legs or back. He will nearly always drop his gun, and I would give him a crack or two more if he needs it."

Trying to run the lab with very little money, Zetek refused to let visiting scientists off the island and then charged them for the extra days they stayed. They complained, but Zetek had firm control of the boats.

And so the scientists, who recognized Zetek as an expert on termites but little else, came to undermine his authority. They'd surreptitiously shoot pygmy mice in the clearing; they mocked his use of the motorized lift.

For Enders, mounting the steep steps to the lab was a physical test. "Anytime I can't make that group of steps in one verse,

I'll stop coming down," he said. Eventually the scientists developed a habit of being away on days when Zetek visited.

As I learned BCI's history, I also learned, little by little, what the current residents were up to. I eavesdropped at the dining table, I attended Bambis, I buttonholed researchers outside their labs. David Marsh was leaving the island for a while, and I asked if he needed help packing up. We went into the forest one afternoon to retrieve a dozen breeding ponds.

Marsh was studying territoriality and mating patterns among *Physalaemus pustulosus*, tungara frogs. These brown creatures the size of a grape aggregated at night to call for mates. Marsh wanted to know exactly how far the frogs would go to aggregate, and if some did disperse, how far they would travel.

Previous frog work on BCI made it easy for Marsh to set up his study. He knew where and when to find the tungaras, and he knew they'd be attracted to his breeding ponds, which were actually plastic dogfood bowls, to sing in a chorus for females. They sang because the females wanted them to; it gave them a criterion for choosing mates. "Like a singles bar," Marsh said.

Nearly every night Marsh surveyed the pond arrays he'd set out in the forest. Wearing a headlamp and carrying scissors to mark the frogs' toes, he counted the courting males and calculated how far each had traveled to sing. The frogs, he discovered, had no loyalty to a particular breeding pond. They didn't defend a territory because the females didn't care. What females "selected for" — cared about — was a long call, which researchers had recently found to be correlated with a more physically fit frog.

The project seemed a little arcane to me, but I was getting used to this. I'd recently had a long conversation with Sabine Stuntz, a German doctoral candidate who was counting the arthropods that live among three kinds of epiphytes on the branches of one kind of tree. She explained her work clearly, but an hour later I couldn't remember why it was so important — either to our understanding of the forest or to conserving it.

In the months to come I'd meet one professor who studied whether white-faced monkeys used their left hand or their right to grab food and another professor who counted the calories consumed by howler monkeys. In a lab upstairs from mine was an entomologist who put miniature bar codes on bees so he could monitor their comings and goings from the hive. Down the hall from him was an ecologist who tracked the demographics of seven thousand spiny rats on twelve nearby islands. A few doors farther along was a woman who performed cosmetic sex-change operations on damselflies, coloring the wing spots of males to make them appear female. She wanted to know if females responded to males' appearance or to some other cue.

Walking the forest with these researchers or their assistants, I would become completely caught up in the excitement of their chase for answers. But at the end of the day, alone in my room, I'd have to ask myself what it all meant. I wondered if the grad students were merely trying to prove that they knew how to set up and run an experiment, if the tenured professors were merely collecting grants to justify their salaries.

BCI's scientists had worked for years to learn that the island contains 270 tons of wood per hectare; that its trees produce 7 tons of dry-weight leaf litter per hectare per year; that leaves, over the course of their lifetime, lose 8.3 percent of their area to insects. But how could such tiny puzzle pieces of information matter? And who would fit those tiny pieces together?

Very few people, it seemed, were asking big questions. Was it because they'd already been answered, or because they'd fractured into ever more numerous, smaller questions? Tropical ecosystems were intensely complicated places. And the closer tropical biologists looked at them, the more complicated they seemed. Would the puzzle, I wondered, ever be complete?

*

By the 1930s, visiting scientists were publishing papers on BCI's insects, birds, rocks, flowering plants, and snakes. The papers were largely descriptive. Experimentation wasn't in vogue, and much of animal behavior was unknown. The field was wide open. Newspaper and magazine writers visited; Frank Chapman's 1929 book about BCI, *My Tropical Air Castle*, became a popular success. The island's reputation flourished, living conditions improved, the trail system was enlarged and mapped.

From its inception and into the 1960s, BCI was, theoretically, supported by table subscriptions — annual fees of $300 paid by each of a dozen institutions to lease lab space. But the system never covered all the station's costs, and subscriptions often went unpaid. Barbour, Fairchild, and Zetek made constant appeals to friends, institutions, and the government for funds. Between 1924 and 1940, the lab survived on little more than $6,000 a year.

The financial situation became especially precarious during the Depression. Zetek let staff go, and Barbour called for continual belt-tightening. It wasn't cheap to get to BCI, and the American research community wasn't especially interested in tropical biology. The station would probably have collapsed entirely if it hadn't been for Barbour's and Allison Armour's repeated infusions of personal cash.

In 1940 Barbour went before Congress to ask the government to assume administrative control of BCI. He wanted the lab to have the protection of a permanent governing body, and he wanted to establish an affiliation that would make potential donors more interested in the station. He requested $10,000 to hire a permanent director, strengthen infrastructure, and purchase apparatus.

By this time there had been more than four hundred papers published on investigations made on BCI, some of them quite well known. C. Ray Carpenter's exhaustive study of howler monkeys in the wild was the first of its kind. Thomas C. Schneirla was well into his thirty-five-year study of the ecology and behavior of army ants, which would become a classic of pre–World War II animal behavior.

Anticipating public skepticism of such basic research, Schuyler Otis Bland, the chairman of the Committee on Merchant Marine and Fisheries, asked Barbour, "What is the practical benefit of the scientific studies you are making?"

"The study of termite resistance," Barbour promptly answered. He had more: the study of avian malaria and parasites in monkeys was important, he said, as these animals served as possible reservoirs of diseases to man.

Malaria and yellow fever were not far from Americans' minds. During the French effort to build the Panama Canal, between 22,000 and 33,000 men died, mostly from these diseases. The Americans wouldn't even begin to dig until the U.S. Army's Dr. William Crawford Gorgas eradicated mosquitoes from the Canal Zone.

James Zetek presciently played the conservation card. "Consider," he said to the committee, "that at the rate the native destroys forests, that in about 25 years Barro Colorado Island will be the only remaining sample of the humid tropics in virgin state. We must maintain it."

John E. Graf, the associate director of the Smithsonian Institution, a BCI subscriber, said, "This island is in the tropical fringe that surrounds us, and it is very important to us to know something about the animal and plant life there."

Graf's declaration would be the motivating force of Barro Colorado's scientists for years to come: learning about the world around us. Panama *is* on the fringe of Meso-America, and the Canal Zone until recently was the only example of continental tropics that the United States controlled. Fifty-one percent of all North American migrant bird species — including many kinds of protected waterfowl — fly to wintering areas in the Neotropics. Panama's birds, then, are also *our* birds.

And so are many of its mammals, who three million years ago traversed the land bridge between the South and North

American continents. Some of our mammals have antecedents in the Neotropics, and some of them have descendants there as well. To know the flora and fauna of the isthmus is to know more about the flora and fauna of the United States.

Researchers on BCI are still comparing temperate and tropical life forms. David Fairchild, in 1940, declared such comparisons "crucial" to developing new perspectives. But while scientists in his day collected very basic information about individual organisms, scientists today are starting to get a handle on how those organisms fit together. If the earlier era was about content, then the modern era is about context.

Congress eventually decided to award BCI $10,000 a year. The money put up another lab building and a larger-capacity water tank. In 1946 Congress gave BCI a permanent steward: the Smithsonian Institution.

The connection sounded grand, but the Smithsonian was at the time a small and overextended establishment. And so the lab in the jungle continued to teeter on the edge of poverty. Researchers still paid just four dollars a night, and they still had no steady source of electricity or hot water. Many scientists were reluctant to stay long under such conditions, especially if they'd brought along wives and children. For days and weeks at a time there was often just one person at dinner.

In an effort to boost interest in the tropics, Robert Enders, who was often that lone diner during the postwar period, began bringing down groups of undergraduates, who had not been welcome on the island in earlier decades. The undergrads kept Enders company, and they kept the data, collected since BCI's founders set foot on the island, flowing.

By the end of my first week on the island, I'd made some progress in habituating my study subjects to me. A field assistant named Paul leapt on my offer of help; soon we'd be censusing spiny rats on several nearby islands. Bret, who'd been so skittish with me, visited my lab with a bat in his hand — a winged peace offering, with thumb-claws and teeth. Chrissy invited me into the forest. She didn't need help with the monkeys, because she couldn't find her troop, but she seemed to want some company.

If my days were regularizing, my evenings were still undefined. No one had invited me to the bar yet, and I felt funny about showing up there alone. I still wasn't sure of my social status on the island. But I could wait. I was gleaning data, and there would be plenty of time to make friends in the months to come.

For now, I spent my leisure time getting to know Bert Leigh, in his apartment up the hill. There, I'd take refuge from the residents, sip a civilized bourbon, and listen to stories about the old days.

3

Cocktails with Bert

BERT LIVED half the week on BCI, in an apartment that abutted the laundry room, and half the week in Gamboa, with his wife, Lizzie. They had two grown children, whom they'd raised in the now-derelict Yellow House, a two-story stucco dorm at the top of the old lab clearing. The residence was informally known as Toad Hall, which made Bert, to himself at least, "Toad."

A reverent, cerebral man with a Princeton Ph.D., Bert has a reputation that looms large among evolutionary biologists. In *Naturalist*, E. O. Wilson includes Bert in his list of the forty "best ecologists and population biologists in the world to the doctoral level." "The legendary Bert," wrote Margaret Lowman in *Life in the Treetops*, "is a brilliant biologist who can expound on almost any scientific topic and is very endearing in his mannerisms, including his predilection to drink scotch with visiting scientists."

Bert invited me up for drinks my first night on the island and then about every week or two thereafter. In the beginning, it was a symbiotic affair: I wanted information about the field station and the forest; he wanted information about the residents. What was Bret doing with his time? How were people reacting to me? Bert was my bridge to the past; I was his conduit to gossip. Soon, though, our relationship would become lopsided. At Bert's invitation, I asked a lot of pesky questions and he, generously, answered them. In ecological terms, we became commensalists: I derived benefit from my host while doing him no harm.

"Come in!" he shouted that first evening when I knocked on his door. Baroque music boomed from a portable CD player. Bert didn't smile; he barely looked at me. "Sit where you'd like." His voice was a little whiny, but that was part of Bert's cultivated eccentricity — a seeming impatience for ordinary social routine.

A book-cluttered coffee table separated the couch, against the back wall, from two chairs. The furniture was upholstered in a yellowy-orange floral fabric, stained and shiny with decades of use. Bert sat on the couch, as he always did when he had a visitor, I would learn.

"What would you like? I've got whiskey, rum, and perhaps a bottle of sherry." The offer was made almost grudgingly, but again, that was part of Bert's style.

While my host poured, I scanned the wall of books that divided the living room from his bedroom. Hardcover tomes, tropical biology mostly. Books on insects, water, lizards, plants, soil, math. I had the feeling he'd read every one. On shelves by the door and by his office, he kept his fiction, religion, and philosophy. The apartment had no kitchen. Essentially, it was a bachelor pad.

Bert looked older than his fifty-eight years, on account of his bristling white beard, his black-rimmed glasses, and a certain frailty of movement. His long-sleeved shirts were always buttoned at the cuffs and neck, whether he was working in his office or walking in the woods. Serious about religion and serious about evolutionary theory, he attended church every Sunday in Gamboa, sat through Bambis every Wednesday and spent Tuesdays at the Smithsonian's Panama City headquarters, where he hurried between his office and journal stacks in the library.

Bert published slowly and steadily, synthesizing the work of ecologists and other scientists. He wrote on the major transitions of evolution, on the evolution of mutualisms, and on the role of selection in evolutionary theory. "Beneath the mathematical wizardry and the irreverent, often brusque style reside[s] a perceptive, highly informed man with a biological intuition more typical of a field scientist than of a theoretician," wrote Geerat Vermeij of Bert in *Privileged Hands*. "He [thinks] about complexity and stability in thermodynamic terms."

By pouring enough Jack Daniels, Bert had managed to wrest sufficient material from thirty-six scientists to compile *The Ecology of a Tropical Forest*, a comprehensive primer to the ecology of BCI, a touchstone for anyone who visited, or hoped to. While I was on the island, he was finishing up *Tropical Forest Ecology: A View from Barro Colorado Island*. Whereas the first book emphasized seasonal rhythms and their role in regulating plant and animal populations, the second book emphasized the role of mutualisms, or cooperation, in shaping the tropical forest. The book acknowledges competition as a primary shaper of organisms and their communities but argues that biological understanding must also reckon with interdependency, the role that mutualism plays in ecological organization, if there is to be any hope for preservation.

My conversations with Bert were often one-sided: given the slightest opening, he'd propound on a topic in which I had no fluency or gallop away on a long mathematical tangent. I didn't mind. I liked Bert's voice. It was musical. He used interesting words, and after years in the jungle he had perfected his own alien accent. "House," on Bert's tongue, rhymed with "mice." He said, "If you start to worry abite that, you're likely to find yourself in queer stirrups."

For a man of Victorian-seeming sensibilities from a socially prominent Virginia family, Bert's conversation was oddly ejaculatory. He was confounded by a phenomenon in "the *damn* forest." He'd ask, "Now what in *hell* is that insect there?" Walking me through a calculation for quantifying herbivory, he'd say, "Then you multiply all this *crap* by 365." Bert abjured eye contact and rarely allowed himself a smile. I felt like I'd do anything to get him to laugh.

"When did you first get to BCI?" I asked as we settled back with our drinks.

"I arrived at Christmas of 1966," he said. "I was a grad student, and I knew very little about the tropics, or BCI. I knew I wouldn't have to cook for myself."

It took Bert, shy and not fond of physical exertion, several days to find his way into the forest. He lived among a handful of scientists in the bunkhouse above the old dining hall. When summer came around and the island grew crowded, he roomed with a young grad student named Robin Foster in a dorm near the docks, long since demolished.

I asked Bert what he did in those early days. "I was a goddamn theoretical biologist," he said. "I gave a seminar on why, when an allele can spread itself by biasing meiosis in its own favor, meiosis is generally fair." I nodded, feeling chastised.

"There weren't any field assistants then," Bert continued. "The Hladiks, who were French grad students, were studying monkey diet and food supply. They hypothesized the monkeys were eating one kilogram of food per day, half of that fruit. That's within ten percent of what we know now." Bert paused to rate my reaction. I knew that accurately determining a ten percent differential, in ecology, was pretty impressive.

That year, the Smithsonian had reorganized its operations in Panama as the Smithsonian Tropical Research Institute (STRI). The island's manager was Martin Moynihan, who had arrived on BCI in September 1957, after Carl Koford — who followed James Zetek as lab director — was fired. Koford, Bert gave me to understand, was but a blip in BCI's history. It was Moynihan, a handsome young animal behaviorist, who would leave his mark on this place, who would transform what had essentially been a nature preserve into one of the most respected laboratories for tropical research in the world.

Bert rose from the couch and went to his bedroom, which I

could see contained one twin bed, neatly made, and a lot of books. I furtively lowered the music by just a notch. When he sat down again, he held a bar of dark chocolate — Swiss, with nuts. He cracked off a small piece for himself and invited me to do the same. With the music quieter I found it easier to talk, but Bert didn't seem to notice.

Moynihan came to Panama with great credentials, Bert told me — he had degrees from Princeton and Oxford universities, and he had studied under the Nobel Prize–winning ethologist Niko Tinbergen and the evolutionist Ernst Mayr, with whom he had written a paper at the age of eighteen.

"Martin was like an eighteenth-century French aristocrat," Bert said, leaning back and crossing his legs. "He was interested in everything, and he had an unquestioned sense of his own authority that he apparently was born with."

Charismatic and energetic, Moynihan began to attract a distinguished clientele to the jungle island, which he fondly called "the green hell." (He may have been quoting U.S. Navy Lieutenant Isaac Stern, who called Panama's Darién region the green hell after he almost died crossing it in 1854. Or Moynihan may have come up with the expression on his own — an example of convergent evolution. Bert used the expression as well — in homage, I suspected.)

Moynihan didn't care whether his scientists were interested in bats or rats or flower buds. "He simply wanted intelligent researchers interested in studying evolutionary problems," Bert said. Moynihan had studied gulls and mixed-species bird assemblages, but his particular interest on BCI was the social behavior and communication of howler monkeys.

Wearing sneakers and clenching a pipe between his teeth, he studied the animals in the forest and in cages that he built in the dank woods along Allee Creek. He ran some experiments, ascertaining which leaves monkeys preferentially ate, but Moynihan's work, like that of his contemporaries, was more qualitatively descriptive than quantitative. It would be up to the primatologist Katie Milton, in another decade or so, to measure the exact caloric content of monkey food.

Moynihan and the men he brought to Panama were among the first generation of students to be fully exposed and committed to the modern evolutionary synthesis, of which Ernst Mayr had been a principal architect. The synthesis combined Darwin's theory of adaptation arising through selection of advantageous traits with the principles of modern genetics, which supplied an explanation (that is, mutation, genetic drift, and genetic recombination) for variation within a species.

Informed by these ideas, Moynihan's team turned its attention to parsing symbiotic relationships between species, complex evolutionary adaptations, and the social behavior of animals. He was, in effect, anticipating the coming recrudescence of interest in tropical ecosystems. "Martin saw the need to know before anyone else did," Bert told me. "He had a vision."

In July 1968, just a year and a half after Bert landed on Barro Colorado, Moynihan shipped him away. He had recognized in the awkward young man a precocious talent, and he had asked him to compare the structure and physiognomy of tropical forests in different regions around the world. Moynihan was interested in knowing how representative the ecology of Barro Colorado's forest and the behavior of its animals were of tropical forests in general. Bert wouldn't return to BCI until January 1969, when Moynihan appointed him to the STRI staff.

Bert leaned forward to refill our glasses. He wiggled his foot, delicate in its tightly laced army boot, and tugged at his ear. "Before Moynihan arrived," he said, "BCI's scientists weren't making connections between fruit, population changes, rainfall, and animal behavior." Residents worked alone. The lab had no coherent research program; nor was one desired. "Throughout the sixties, people were still doing their own thing," he said. "But by the seventies, you could see the puzzle pieces fitting into the picture, without anyone micromanaging it. The seventies were the Golden Age of BCI. This was *the* central place to study wild Neotropical mammals. The station, in the summer, was jammed with researchers."

For the first time in BCI's history, lab residents shared an intellectual bond, and a distinct theoretical orientation began to emerge. The work that came out of the lab, said Bert, "was biologically significant, and exciting." The research program attracted attention and funding. The number of visitors mushroomed.

Bert nodded toward my nearly empty glass. "Help yourself," he said, pushing the bottle toward me. I glanced at my watch. Ten-thirty! Bert showed no sign of fading, but I was ready for bed. I finished my drink and thanked him for his time. Then we arranged to meet in the morning for a short forest tour. When I opened the door to leave, Bert said, "Be of good cheer."

With his collar tightly buttoned and his sleeves rolled down, Bert led me at a stately pace through the old lab clearing, into Lutz Ravine, and up — very slowly — the 45-meter (137-foot) tower, built by STRI to provide access to the canopy and for collecting meteorological data. The tower consists of narrow platforms connected by slanted ladders. Guy wires hold the whole thing in place.

I'd climbed up the tower on my own a few times. From the very highest level, the view was spectacular — out over Slothia Island in Lab Cove, over the shipping channel, and on to the hazy jungle that stretched to the horizon. The treetops of BCI, between 100 and 150 feet up, formed an undulating green sea, dotted for half the year with flowering emergents. At the end of the dry season, *Tabebuia guayacan* burst into bright yellow flower. From February until May it was *Jacaranda copaia*, which flowers in purple, and from April till June it was the yellowy orange blossoms of *Vochysia ferruginea*.

The general evenness of the canopy was impressive: no tree grew higher than it had to grow to capture the optimal amount of light. At forest edges or in clearings, where trees had greater access to light, trees sent out branches lower than did their conspecifics in the forest interior. A large guanacaste tree well along Wheeler Trail and surrounded by smaller trees had substantial branches ten feet up its trunk — evidence that the area around it had once been cleared.

It made me a little dizzy to look down into the grove around the tower, where vines and epiphytes scrabbled one atop the other. As often as I tried, I could never follow one tree from its base to its canopy. A riot of green interlopers — smaller trees, orchids, ferns, mosses, and bromeliads, the stuff that everyone imagines when they imagine a tropical rain forest — always led me astray.

The forest canopy has been so little studied that scientists call it the final biotic frontier. Intrigued by this place he could not reach, Alexander von Humboldt described the canopy, in 1815, as "a forest above a forest." In 1917 William Beebe wrote, in *Tropical Wild Life in British Guiana*, "Yet another continent of life remains to be discovered, not upon the earth, but one to two hundred feet above it, extending over thousands of square miles."

This new continent, because it received a great deal of light and had a high productivity level, attracted a multitude of organisms. The majority of the world's insects, the Smithsonian's Terry Erwin has suggested, live in the canopy. Yet it wasn't until the 1980s that researchers began to explore the canopy using walkways, webs, trams, cranes, balloons, and single-rope climbing techniques. As went the evolution of climbing technology, so went the evolution of canopy studies.

In 1992, STRI erected a forty-meter-tall canopy crane, with a thirty-five-meter arm, in Panama City's Parque Metropolitano, which was followed by five others around the world. The crane allowed scientists to study the behavior of arboreal mammals, insects, and birds, the chemistry of leaf tissue, and the mechanisms of water transport and stress. They could compare the forest at different heights. Biologists had known for a long time how fast small things grow, but only when the towers went up did researchers turn their attention to how fast big things, like hundred-foot trees, grow. Now they could measure herbivory in the entire forest rather than in only the five percent they could reach from the ground.

The lowest levels of the tower, below most of the clutter of vegetable matter, also made a convenient spot for animal watching. I once sat here just before the tropical equivalent of dusk — which is to say, the precious last moments before everything green slams straight into black — and spotted a pair of brocket, tiptoeing delicately into Lutz Stream. These small deer, no more than a meter high at the shoulder, are Central American representatives of the family Cervidae, in the genus *Mazama*. White-tailed deer, in the genus *Odocoileus*, also bound around the island, which surprises a lot of day visitors who have settled into a tropical frame of mind and are looking instead for monkeys, anteaters, sloths, and blue morpho butterflies.

They'd do well to remember that Barro Colorado's current roster of plants and animals stems from the mixture of two very different biotas. Their conduit was the Isthmus of Panama, which three million years ago began to rise from the sea, serving as a land bridge between the hemispheres. The corridor allowed animals we think of as northern — deer and raccoons — to wander south, and such "southern" animals as opossums and sloths to wander north.

For most of the Age of Mammals, which began 65 million years ago, North and South America existed in isolation from each other, and so evolved very different life forms. The South American bestiary included carnivorous marsupials, herbivorous ground sloths that stood more than six meters high on their bearlike legs, glyptodonts, which resembled VW-sized armadillos, and three-meter-long, rhinolike toxodonts, considered by Darwin to be "among the strangest animals that ever lived." There were rabbity typotheres, long-trunked pyrotheres, and gomphotheres with short trunks, giant molars, and tusks growing both down and up.

Regular interchange with the Old World over the Bering Strait gave our North American mammals an obvious resemblance to those of Eurasia and Africa. After the Panamanian isthmus arose, North American cats, dogs, bears, mastodons, and deer headed south, while South American armadillos, porcupines, and opossums migrated north.

The land bridge wasn't always a reliable conduit: two or three times since the beginning of the Quaternary, it sank beneath the sea, leaving its mountaintops to stand as an island chain for many thousands of years. During those periods, when animals could not cross between the hemispheres, they grew to be less and less alike. The Cervidae branched and produced brocket; the Procyonidae produced the kinkajou. In North America the camel family died out, while in South America it produced the llama and the guanaco. Lemurs and tarsiers died out in North America but persisted in the Southern Hemisphere. Rodents all over South America flourished, and Panama fairly bristled with porcupines, guinea pigs, viscachas, agoutis, and capybaras, aquatic creatures that grow up to one and a half meters long.

The largest of the native South American mammals suffered an extinction about 12,000 years ago, as the North American fauna got a foothold. The isthmus was last under water many thousands of years ago. Since then, the land has been rising at a rate of one meter a century.

Several platforms shy of the tower's roof, Bert and I stopped. "I think this is high enough for our purposes," he said. As he caught his breath, he added, "This is not a particularly odious place for watching morphos." We watched two flit about at different strata of the forest, and then the conversation turned to architecture.

"That branch is distichous," Bert said, pointing out horizontal twigs with leaves either alternate or opposite. "This one is decussate" — its leaves were arranged in pairs at ninety-degree angles to the pair below them. Hungry to learn the scientific names of plants and animals, their habits and haunts, I took notes and sketched.

With his esoteric knowledge and intense focus, Bert at first intimidated me. As time passed, those feelings turned to awe. Bert was a master, with a firm grasp on so many tiny aspects of the forest. Whereas I was content to tiptoe the path of a generalist, learning a little here and a little there, Bert, with his constant input, seemed to expect slightly more diligence from me.

"You see that *Virola*," Bert said, pointing to a tall tree behind the tower. "It's lost its apical dominance. Its crown was infested by lianas and its trunk now forks out several times." My comprehension was sometimes spotty. I noted the differences in leaf design and branching patterns, but I didn't understand how one could be better than another. It was Bert who started me thinking about the processes rather than the results.

If I had been in the forest with Frank Chapman, one of BCI's early naturalists, I would have had the benefit of his keen observation and enthusiasm. But Chapman and his cohort didn't have nearly the detailed ecological knowledge or perspective of today's residents — Bert, in particular. Where BCI's early scientists saw trees as competing individuals, Bert was more inclined to consider them members of an interdependent community.

In his latest book Bert had asked what comparisons of the spectrum of architectures among different forest strata and among different forests can tell us about the meaning of a tree's leaf arrangement and branching pattern. The trees of lowland tropical forests look remarkably different from each other, Bert noted, yet they share many important features. Why?

Trees, it was now established, arrange their leaves according to such factors as their growth rate, their photosynthetic needs, the soil they sit in, and the surrounding forest's structure and climate. A tree can be quantified, after all, and once quantified, retrodicted. If we could predict the architecture of trees in different biogeographic realms, Bert taught me, then we'd know something about the predictability of evolution itself.

I talked frequently to Bert about Martin Moynihan in the months to come, and he always stuck to the high road. But I would glean from others that Martin, although largely responsible for shaping the island's intellectual history, was a difficult person, idiosyncratic and arrogant. He didn't want to spend money maintaining the shelters at trail ends, so he abandoned them to rot. He didn't want to worry about the herbarium's thousands of delicate samples. One day, in a fit of temper, he simply threw all the voucher specimens away. "He didn't believe in collections," Joe Wright, a craggy-faced STRI staff scientist, told me. "He thought they were a ticket to doom."

Neither did Moynihan believe in having a washer and dryer on BCI, which was a sore point with any resident who did fieldwork wearing yesterday's wet, dirty clothes. Relying on the Canal Zone laundry service meant, Robert Enders once calculated, that scientists needed approximately fourteen changes of clothes.

A visiting primatologist condemned Moynihan as "an insane alcoholic with a terrible temper." Even those who worked with him on a daily basis feared his outbursts. A bad idea or wrong thinking could send him into a rage. Once he flung a plate against the wall of the dining hall. Scientists whom he didn't invite to work on Barro Colorado were especially hard on him. Still, the force of his intellect and his personality continued to draw topnotch scientists to the island and to his swank dining room in Panama City, where the wine, not unlike Bert's liquor, freely flowed.

The progress of tropical biology as practiced on BCI mirrored the progress of biology in general. Early naturalists like BCI's

Barbour, Wheeler, and Chapman collected organisms and described animal behavior. They were working a blank canvas, roughing out large swaths but leaving out, for the time being, much detail. Their collections and observations, and those of the nineteenth-century naturalists who came before them, form the foundation of modern tropical biology.

As knowledge accumulated, scientists began to search for the underlying principles that organize nature. Moynihan arrived in Panama on the cusp of this change. He was the evolutionary link between the describers of the past and the quantifiers of the future. The quantifiers would make use of statistical analysis, rigorous testing of hypotheses, and experimental manipulation of natural systems.

Moynihan himself was not a rigorous scientist. He discouraged ecologists whom he deemed to have narrow research approaches, and he hired animal behaviorists. But times were changing. More and more plant people were showing up on BCI. The rigor quotient was about to go sky-high.

"Vertebrate studies were going out of fashion," Bert said to me. We were clambering down from the tower, one slow flight at a time. "They faded because hypotheses about animal behavior aren't as testable as standard ecological hypotheses." Resting on a platform, he raised his eyebrows and we made eye contact for a split second. "There's an idea that studying an animal will change it." He turned abruptly and headed down another flight.

In 1967, Robin Foster arrived to study the island's seasonal rhythms of fruiting and leaf flush. Using Paul Standley's *Flora*, a 1927 field guide, Foster, a grad student from Duke University — and Bert's roommate at BCI — taught himself to identify fruits and flowers. Later he'd learn the leaves of the island's thirteen hundred plant species, a daunting endeavor; very few people in the world can confidently identify most of the trees in a plot of Neotropical rain forest. Over time, Foster would develop a herbarium so that residents could key out even those plants that lacked flowers. By teaching fellow researchers to identify plants, he'd shape the direction of research on BCI for years to come.

On Moynihan's watch, the Smithsonian's operations in Panama grew logarithmically. Russia's launch of Sputnik, in 1957, had sparked a new, competitive emphasis on scientific research in the United States, which extended even to its tropical fringe. Between 1960 and 1970, the Smithsonian's yearly budget increased more than sixfold, to $49 million. Money from the Smithsonian, the armed forces, and the National Science Foundation began to trickle down to the laboratory in the jungle.

In 1965, BCI established radio contact with the mainland, and the NSF awarded the station a $100,000 grant, which paid for a power cable to be laid under Gatun Lake. For the first time the island had a reliable supply of electricity, and almost immediately the scope of research blossomed. Scientists who needed freezers and drying chambers to preserve samples arrived, as did animal behaviorists with electronic monitoring equipment. The novel combination of nearly unspoiled jungle with modern laboratory equipment further raised BCI's currency.

By 1970 the balance of plant and animal people on Barro Colorado had begun to tip: Robin Foster's influence was spreading. Moynihan moved off the island and, in 1974, resigned. Ira Rubinoff, a marine biologist, became STRI's new director. Evolutionists, who explore the mechanisms that influence variation and change in organisms and groups, were giving way to ecologists, who study living organisms and their interrelationships with the physical community. Their work, Rubinoff correctly assumed, would become increasingly important as scientists addressed issues of global change.

Unlike the naturalists who'd surveyed this region centuries ago, moving quickly over thousands of square miles, the ecologists of the 1970s and '80s focused intensively on individual sites, asking how the forest functions and maintains itself. Long-term studies came into vogue. On BCI, STRI established the Environmental Sciences Program to investigate whether the tropics were as stable as theorists had suggested. The essence of the program was data collection. As ever, technicians tracked meteorological and hydrological conditions, but now they also monitored populations of mammals, lizards, and insects. They recorded the times of leafing, fruiting, and flowering in selected trees.

Visiting biologists were now studying plant pollination, seed dispersal, seedling establishment, seasonality, herbivory, pathogens, photosynthesis, water relations, and tree growth and death. Faster, more powerful computers let them collect and analyze enormous quantities of data. Advances in miniaturization let behaviorists — there were still a few around — track free-ranging animals, and the latest techniques of molecular biology allowed evolutionists to re-address old questions. DNA sequencing, for example, revealed genetic relationships among organisms that earlier researchers could only guess at.

Moynihan, who had paved the way for this modern era by setting the research station's intellectual tone, remained on staff after resigning as director. When he arrived on BCI in 1957, he was the island's only resident biologist. By the time he picked Rubinoff to replace him, he had hired nine permanent staff biologists — Bert Leigh and Robin Foster among them — and instituted a fellowship program for grad students and postdocs that continues to this day.

As soon as Moynihan moved off-island — to study squid communication in the Pacific — a washer and dryer arrived on BCI. According to Robert Enders, it was the greatest advance in the research station's history.

As Moynihan was settling into new digs at STRI's marine station, all hell was breaking loose on BCI. Famine had settled over the island. Because the dry season of 1970 had been very wet, only one-third the normal amount of fruit fell in the forest between September 1970 and February 1971. "The island stank," Bert told me. We were heading out of Lutz Ravine now, looking at the branching pattern of an *Anacardium excelsum*. "Animals were dying all over the place and the vultures couldn't keep up with the dead frugivores."

As the famine progressed, it became easier and easier to see mammals in the forest: they sought food even at midday and during heavy rains. Trail walkers found at least one dead animal every three hundred meters, even well away from the lab clearing. Animals normally wary of humans — collared peccaries, tapirs, and kinkajous — began to visit the lab clearing in increasing numbers. I turned, later, to Bert's *Ecology of a Tropical Forest* for a description of that period: "The spider monkeys . . . launched an all-out assault on food resources inside the buildings, learning for the first time to open doors and make quick forays to the dining room table, where they sought bread and bananas, ignoring the meat, potatoes, and canned fruit cocktail, and brushing aside the startled biologists at their dinner."

The crisis for fruit eaters was an opportunity for Robin Foster, author of that description in Bert's book. For two years he'd been collecting, identifying, and analyzing nearly everything that fell from trees on the island's plateau. The famine was a less-than-subtle reminder to researchers that the forest knew nothing of the status quo, and it spurred Foster and others to ask what role seasonal shortages of leaves and fruit played in regulating animal populations.

One study showed that fruit shortages left howler monkeys in poor shape, unable to resist botflies. Screw worms then colonized the botfly wounds, and the monkeys fell out of the trees, dead. Another study demonstrated how agoutis, weakened by fruit shortages, were taken out by coatimundi. In times of plenty, these animals would feed side by side.

The more these scientists looked at seasonal shortages, the more their view of life on BCI changed. In temperate-zone forests, predators controlled vertebrate populations; here, where large predators had nearly been poached from existence, it appeared that plants were calling the shots.

"Foster started a cottage industry in animal-food relationships that's continued to this day," Bert said. We were walking down Wheeler Trail, taking the long way back to the lab. As we passed through a gap, the heat of the day enveloped us.

Suddenly Bert stopped. "This is not an overjoyed forest at the moment." He rested his hand on the bark of a *Socratea*, a slim palm tree supported by prop roots. "It's lost its air conditioning." He looked toward the canopy, which, by evapotranspiring water, manufactured cooler air.

Last year's El Niño had starved the region of rain, and this year's dry season, which we were near the end of, was the driest on record. Some researchers predicted a famine later in the year.

Bert continued down the trail. "Once visiting scientists had a good idea of what fell where and when and which foods animals were eating," he said, "they could pose a whole new set of questions." For example, what accounted for plant diversity on the island's plateau? How did herbivore and frugivore populations regulate plants, or vice versa? How did food availability affect dietary shifts, seasonal birth patterns, and death rates?

The questions, it seemed, were unending; every new bit of information begat another question. Answers could be years or even decades in coming. And then another problem arose: Would findings made on BCI pertain only to this forest, or would they hold true for other rain forests as well? The possibilities made my head spin.

We arrived in the lab clearing, where Bert and I parted company. "Thanks for the walk," I said.

"Be of good cheer," he said, then disappeared up the concrete path.

*

For all the baseline data collected in BCI's forest, science still had an insufficient understanding of the differences between trees and how so many plant species could coexist. Did forests have some sort of collective stability, or were the trees in a state of nonequilibrium? That is, would the forest of the future look much different from the forest of today?

Robin Foster, who'd acted as midwife to BCI's era of ecological studies, was prepared to tackle these questions. In 1980 he set up, with his colleague Stephen Hubbell, the Tropical Forest Dynamics Plot, a fifty-hectare study area (half of a square kilometer) that hadn't been cut, they believed, for two thousand years. The plot, on the island's plateau, would become one of BCI's most famous studies.

With a phalanx of assistants, Foster and Hubbell tagged, measured, identified, and then mapped more than 250,000 stems with a diameter greater than one centimeter at breast height. It was the largest mapping project in the history of tropical biology, and it took a very long time. Every five years researchers recensused the plot, correlating tree growth to such variables as soil, moisture, light, and proximity to other trees.

For all its significance, the fifty-hectare plot is an unassuming place at first glance. Here, as on the rest of BCI, no single tree species dominates, or even comes close. Hubbell likes to say that the biology of the tropics is about the biology of being rare.

Closer inspection of the plot reveals an obsession with measurement run amok: orange flagging, with geographical coordinates, hangs at one-meter intervals; iron candy canes poke from leaf litter bearing even more coordinates. Seed traps stand between buttress roots. Metal pie tins adorn others: the tins reflect the headlamps of researchers who work in the dark, measuring water loss before daylight cues the leaves' stomata — structures through which gases are exchanged with the atmosphere — to shut down.

The plot was proof positive that science is fed by baseline data. Anyone could download the plot's data set and correlate it with their own new measurements. A postdoc, for example, might measure nutrients across the plot, then correlate those numbers with tree growth within and across species.

Over time, plot data would spur studies on the role of light gaps, on competition and pathogens, on the reproduction of tropical trees and on their phenology — the seasonal rhythm of fruiting and leafing. The plot would also inspire twelve other fifty-hectare comparison plots around the world. But all that was well down the road.

First, there was the original plot map to analyze. Its results stunned Foster and Hubbell. Almost every tree species was aggregated, but species were aggregated randomly. Foster, who is now at Chicago's Field Museum, and Hubbell, who teaches at the University of Georgia, had no explanation for this phenomenon.

In the beginning, the team found both equilibrium *and* non-equilibrium forces at work. Then, upon learning that the plot's oldest trees were no more than five hundred years old and that indigenous people had farmed the area for thousands of years before that, Foster and Hubbell realized that the forest couldn't possibly be in equilibrium. These trees were successional — in the process of giving way to an entirely new forest composition.

Science knows almost nothing about tropical forest, Foster and Hubbell admit, and it may take another several hundred years before its intricacies are fully quantified. The time frame is, needless to say, far longer than the parameters of any scientific grant.

By now, two weeks into my stay, I knew my way around many of the forest trails, around the lab clearing, and around BCI's bar. I'd read here in the afternoon, or I'd join a scientist to talk about work. Before dinner a handful of residents would gather for a soda or beer and gaze out over the cove to the shipping channel. The afternoon boat pulled in at six or so. The view from the balcony gave us our first look at newcomers. The sun dropped quickly over the island's summit, well behind us, and the air grew marginally cooler.

On Barro Colorado, as elsewhere, the history of science had progressed from collection and description to quantification and rigorous comparison. The history of scientists at leisure, though, was largely unchanged. At day's end, the latest crop of biologists still put their boots up, as did Wheeler and Barbour, lifted a drink to their lips, and gossiped about absent colleagues, maligned and missed.

I'd soon feel comfortable enough contributing to this history, but I'd never feel completely at home in the bar. And so I continued to make my pilgrimages up the path, take my place in the floral armchair, and pepper the ever-indulgent Bert with questions.

4

The Covert Troop

MY FIRST DAY OUT with Chrissy Campbell was her tenth day without a study subject. The twenty-one members of her spider-monkey troop had gone missing. And so up and down ravines we bushwhacked, along trails, shortcuts, and streambeds. Lately Chrissy had seen numerous groups of howler monkeys and white-faced monkeys but no *Ateles geoffreyi*, of which BCI has just one troop. In other New World forests, *Ateles* species live in smaller groups and inhabit small ranges. Here, lacking competition, they roamed the entire island, searching for fruit and making Chrissy, in her pursuit of them, miserable.

Now and then Chrissy stopped to stare into the treetops through binoculars, or simply to listen. Hours passed with no sign of a primate. At the top of a steep hill she panted, "I've got to stop smoking."

At twenty-six, pale and sweat-slicked, the aspiring physical anthropologist seemed to lack the heartiness of the prototypical outdoorswoman. She was tired and her feet hurt, and she wasn't slow to admit it.

Spreading a sheet of plastic over a rock to foil ticks and chiggers, Chrissy sat down to eat. She stared quietly at the ground, then suddenly blurted, "Dian Fossey and Jane Goodall really glamorized this field, but it's such hard work." It was *Gorillas in the Mist*, the film about Fossey's work in northwest Rwanda, that inspired Chrissy, as a junior at the University of Canterbury in Christchurch, New Zealand, to pursue primatology. She had even bought the videotape. "The animals were incredible," she told me. "They were so similar to man. Immediately I wanted to study them."

Her sandwich disappeared in four bites. "This work is really lonely," Chrissy said. She unwrapped a PowerBar, melted to its wrapper. "You're out here alone, from six or seven A.M. until five-thirty or six at night." Until recently Chrissy had had an undergraduate assistant to help her with the monkeys and acknowledge her complaints, but the assistant had been put out of commission by flat feet and aggravated bunions. Now Chrissy had me.

Chrissy had been among the first residents to invite me into the field. Although I wasn't much help to her at this point, our conversation at least kept her awake. She'd been depressed lately and sleeping poorly, tormented by visions of unidentified primates in treetops.

On BCI, with no home range, the spider monkeys had no regular sleeping spot. If Chrissy didn't follow her troop until the monkeys bedded down, at six P.M., it was unlikely that she'd know where to find them the next day, at six A.M. And if she happened to oversleep by just a minute or two, she'd almost certainly find herself on walkabout, as she did today, looking for a troop that starts to travel as soon as it rubs the sleep from its simian brown eyes.

Chrissy sighed as she rose to her feet. "It looks like it's going to be another day without data," she said. "Another day without observations, another day without fecal samples."

Unlikely as it seemed, the fecal samples, from the troop's five adult females, could make Chrissy's career. Once analyzed in a lab back in California, they would, Chrissy hoped, tell an important story about spider-monkey reproduction — about female hormones, adult behavior, and the relationship between the two. Without near-constant data collection, though,

Chrissy had nothing. Her field period would be wasted. And so as the tenth day without her troop wore on, the pressure inexorably built.

Even without monkeys to look at, I felt lucky to be starting my fieldwork with Chrissy, who in many ways resembled an old-fashioned naturalist, the type of scientist who founded BCI. Unlike so many of today's residents, Chrissy was not collecting data with computers or measuring an obscure, invisible process. Her work, in this phase, was descriptive: she hiked the forest trails and the wilder places in between equipped only with notebook, binoculars, and collecting bags. She observed closely, and she wrote down what she saw. Bates did it. Spruce did it. Darwin did it, too. It was the essence of field biology.

On Barro Colorado, Chrissy walked in the footsteps of the renowned Stanford biologist C. Ray Carpenter, who worked on BCI sixty-five years ago. Carpenter became the first scientist to observe, over the long term, a New World primate in a naturalistic setting. His howler-monkey study became a classic in the field and established a precedent for the work of Goodall, Fossey, and other primatologists to come.

When Carpenter arrived on Barro Colorado, little was known about howlers, and so he had a clear field in which to operate. In addition to studying their posturing, manipulation, locomotion, and feeding, he also studied their territoriality and nomadism. He studied their organization, group integration, social relations, intragroup behavior, group coordination and control, and the relations of howlers to other animals in the same environment. The enormous project, said Robert M. Yerkes of the Yale Laboratory of Comparative Psychology, was "the first reasonably reliable working analysis of the constitution of social groups in the infra-human primates, and of the relations between the sexes and between mature and immature individuals for monkey or ape."

Chrissy's project was far narrower in scope — she was, after all, an underfunded grad student. But her study had the potential to be quietly groundbreaking as well. If Chrissy found her monkeys and collected enough data, she'd be the first scientist to correlate the sexual behavior of female spider monkeys in the wild with changes in their ovarian cycle, information that would help decode the mysteries of monkey reproduction.

Spider monkeys, like human beings, are covert ovulators: they give no obvious visual sign of being fertile, the way chimpanzees do with their sexual swellings. If researchers don't know when monkeys ovulate, they can't make sense of sexual behavior. "A few studies have noted behaviors that seem to be related to ovulation," Chrissy explained as we reconnoitered Wetmore Trail, "but without independently measuring the concentrations of the female hormones progesterone and estrogen — which show up in feces and in urine — their significance can't be determined."

For example, monkeys copulate, like humans, during and outside of estrus. But how often does copulation coincide with estrus? Who initiates the copulation? Is it a matter of female attractivity (how attractive she is to males), female proceptivity (behavior exhibited by females in order to attract a male), or female receptivity (her acceptance of male sexual overtures)?

"How hormones affect behavior is fascinating to me," Chrissy said, stopping to listen for movement in the treetops. "I want to know when the females are in estrus so that future researchers will recognize it. I want to know how estrus affects the social system."

To most people, studying animal behavior has a whiff of romanticism to it: the researcher pads through the forest with only the most rudimentary of tools. In a high-tech world, we may feel a certain nostalgia for such methods. But the sentiment doesn't bring money to scientists. And so Chrissy, at the conclusion of her field period, would exchange her binoculars for a lab coat and run her fecal samples through a sophisticated series of assays at the University of California, Davis. She'd graph the hormonal cycles over time and plot atop them the various behaviors, coded with numbers, that she'd witnessed in the field.

The scope of Chrissy's project was limited, and the questions she was asking fundamental. Still, she was carving her niche in a crowded field by combining straight observation with high-tech analysis on million-dollar machines. The reductive aspect of the project lent an air of rigor to what would otherwise be a pure animal-behavior study, decidedly out of fashion in these molecular times.

The National Science Foundation had given Chrissy its top award for doctoral candidates, and her advisor expected her to get at least three published papers out of her thesis. The idea of being cited by other scientists was compelling to Chrissy. "There are only about seven dissertations on spider monkeys," she said. "People could, in a few years, be citing Campbell."

I asked Chrissy why the study, if it was such a good idea, hadn't been done sooner. "Because working on monkeys in the wild is hard," she said. The troop moves apart and together over the day, a type of organization called fission-fusion; the females show few obvious signs of reproductive cycling; and their sexual behavior takes place in seclusion from other group members, which makes it nearly impossible to observe all the individuals in a social unit at once.

The afternoon wore on. At a light gap near the beginning of Zetek we stopped again to rest and eat. It was sunny here, and our shirts eventually gave up their dampness. We were surrounded by the sounds of a New Age relaxation tape — trilling warblers, the watery cry of the oropendola (a member of the family Icteridae, which contains the Baltimore oriole), the chatter of white-faced *Cebus* monkeys, and the song of poison-dart frogs.

Arrayed around Chrissy were the contents of her fanny pack: one sandwich, two pieces of fruit, her "pharmacy" (aspirin, Band-Aids), Ziploc bags, masking tape (for removing ticks), two water bottles, a compass, a map, one PowerBar, two pens, and a flashlight. A white hard-hat dangled from a loop on her pack. In the cargo pockets of her field pants were her data book and a walkie-talkie.

Chrissy's kit didn't differ all that much from Carpenter's, sixty-odd years before. But in addition to Carpenter's field glasses, notebook and pencil, compass, camera, canteen, and waterproof bag, he also carried a .32 Colt automatic pistol, a machete, and a snakebite kit with antitoxins. When he planned on collecting specimens, he also carried a shotgun, steel tape, and dissecting instruments.

A tick crawled up Chrissy's pants. "So gross," she said, picking it off. It was then I noticed the chigger bites running up and down her arms. Chrissy's work was filled with hazards. During her first week in Panama she tripped on a liana and cracked her wrist bone. Two months later she was standing under the troop, scribbling in her Rite in the Rain notebook, when a *Virola surinamensis* branch, about five centimeters in diameter, crashed onto her head. After getting stitched up in Panama, she bought the hard-hat.

Despite her Jobian trials, Chrissy enjoyed the forest. Novel insects distracted her. She had favorite trees. This morning, she'd made a route choice based solely on her desire to see a tinamou nest that three days ago held five turquoise eggs. The Tinamidae, a bird group endemic to the Neotropics, were brown, chicken-sized birds with pigeonlike heads. Only the males incubated eggs.

When Chrissy found the nest, on the ground in the interstices of two fig buttresses, only a few blue shards remained. "Coati, probably," Chrissy said. "Stupid place to put a nest, tinamou."

Another sandwich disappeared and Chrissy sighed. I asked if she ever had second thoughts about her thesis. "Yes, definitely," she said quickly. It wasn't the validity of her research she doubted; it was the way her research made her feel. Being alone in the field, day after day, had driven her to tears. Chrissy was a social animal: she missed having lunch with other residents, she resented her early bedtime.

She continued. "My mother, she's a psychotherapist, thinks that when I get back I should attend some sort of rapid stress decompression clinic." Many primatologists, she said, come in from the field depressed and unable to function in the "real" world. Louis Leakey's protégées Dian Fossey and Birute Galdikas weren't the most well-adjusted human beings. They were known as high-strung loners, possessive of their study subject and quick to take offense.

Fossey spent close to fifteen years studying mountain gorillas in northwestern Rwanda. She was superstitious and prone to dark moods, and her field assistants never lasted long. Fiercely protective of the troop she had habituated to humans, she inevitably ran into conflicts with locals. In 1985 she was murdered, the case never solved.

Jane Goodall, another Leakey protégée who also went weeks without finding her chimps, spent a year alone in the Tanzanian field. Had it been any longer, she said, "I might have become a rather strange person." Inanimate objects began to develop their own identities, and she said hello to the stream and to her hut. Touching a tree, she felt she knew the underground roots and pulsing sap within.

Chrissy seemed to be anticipating a mental collapse. "No one back home understands what you've been through, the long days alone, the frustration at not finding the monkeys, the physical rigors," she said. "Adjustment to the world of friends and news, stuff that's not remotely like your monkeys and your jungle, can be jarring."

"Most people who study primates have no siblings," Katie Milton, Chrissy's advisor, would later tell me. "They're isolationist, they're not used to being in a support group. Chrissy is very gregarious, she has a wonderful personality." In short, Milton concluded, "she may be too normal to do this."

Chrissy repacked her gear and we marched on. The sun was nearly at its apex now, but watchless in the forest you'd never know it. The light that hit the ground was dim and green-tinged. There were no shadows to guide us. Only one percent of canopy sunlight reaches the forest floor of Barro Colorado, which is one of the factors that make competition among tree seedlings so fierce, and so interesting to those who study the distribution and survival rates of plants. Less than one percent of seedlings in the rain forest reach adulthood.

Through streams we sloshed and up muddy banks thick with palms we scrambled. Chrissy checked her compass and cursed lianas. I felt, cinematically if not vegetatively, as if I were in Vietnam, following a tough chick in army pants and a helmet, the sleeves of her T-shirt rolled against the heat.

When Chrissy first set foot on BCI, she was terrified of the forest — of the insects, snakes, spiders, and, most acutely, of getting lost. She didn't know how to use a compass, had never worked with a map. And yet it never occurred to Chrissy that she wasn't, either physically or temperamentally, cut out for this line of work.

Unlike most of the grad students and postdocs on BCI, Chrissy had no early yearnings to be a scientist. She never considered herself a naturalist; she wasn't prone to walking in the woods. Her family camped, but it was "yuppie style," she said, and took place in a trailer. An interest in "cute, furry animals" gave her ideas about becoming a vet, but the idea wasn't so compelling that she ruled out becoming a dentist, or even a hairdresser. "I could do that," Chrissy said.

Improbably, Chrissy excelled at high school chemistry and biology. She pursued a zoology degree at university and studied allogrooming among spider monkeys in the local wildlife park for her honors thesis. As a student Chrissy had read the papers of Katie Milton, a well-known Berkeley primatologist. After viewing *Gorillas in the Mist* a few more times, Chrissy wrote to Milton and was accepted into the Ph.D. program. She'd conduct a one-month pilot study on BCI in the spring of 1996 and then return for her actual study in October 1997.

Just past Zetek 5, Chrissy paused and cocked her head. A laughing bark came from the open stands to the north. "Woo-hoo!" Chrissy shouted, literally jumping into the air. "That's them!" She skipped down the trail and plunged into the forest.

"Woo-hoo!" I echoed unselfconsciously, hot on her heels. For some reason, this gleeful expression had infected the island. The vector was youthful fans of *The Simpsons*, who'd spread the expression to young and old, gringo and Latino, researchers and administrators. It had spread, in the lingo of evolutionary biology, as a meme — through cultural transmission.

Chrissy followed the sound for about twenty meters and stopped. Binoculars to her sweating face, she assumed the position: neck craned, feet firmly planted. I tried to do the same, but after ten minutes had to support the back of my head with my hand. "Your neck gets strong real quick," Chrissy said. The monkeys were feeding in the trees all around us. "You bastards," she shouted at them, smiling.

The monkeys looked entirely innocent, not like bastards at all. They were lithe and short-haired, with a reddish cast to the fur on their torsos. Their mouths were neat and expressive, their nostrils delicate, eyes ringed by whitish fur, faces dished. They reached with languid fingers for branches, dangled from their prehensile tails on vines. The monkeys weighed about 6.4 kilograms, with tails nearly a meter long.

This was my first glimpse of a spider monkey, and yet it wasn't the animal's beauty or grace that initially caught my eye. Rather, it was the light-colored, ten- to twelve-centimeter-long appendage between the females' legs.

"That's usually how I recognize them, by the size or color of their clitorises," Chrissy said. She wasn't joking, although she also used differences in hair coloration, scars around eyes, and other facial markings. The pendulous, hypertrophied clitoris was largely a mystery: it didn't seem to play a role in sex, and Chrissy had never observed masturbation. "It may have something to do with leaving scent markings for males," she said.

With sinuous, spidery moves, the monkeys linked and danced through the topography of the treetops. Their locomotion and anatomy were actually quite similar to that of apes, which was one reason they interested primatologists. I watched them leap in silhouette from tree to tree, plunging sometimes seven meters through the air — arms, legs, tail, and clitoris akimbo.

Chauvinistic about her study subject, Chrissy found the spider monkey a far superior animal — smarter, prettier — than the island's roughly twelve hundred howler monkeys, with their black eyes, low foreheads, flat noses, enormous mouths, and scraggly hair. Frank Chapman described an old howler male as the "incarnation of every evil thought that has ever passed through the mind of man."

In contrast, the spiders seemed possessed of a light and playful spirit. Juveniles grappled with one another, mothers dandled babies. There was plenty of fruit to go around. Perhaps it was just Chrissy's relief at finding the troop, or mine at finally seeing her smile, but the forest just then seemed like Eden.

Oddly enough, the spider monkeys in the trees above us were not native to Barro Colorado. They were the third-generation descendants of animals brought to the island, starting in 1959, by Martin Moynihan. Panama's wild primate populations were rapidly disappearing, and Moynihan wanted to restore the island's mammal population to its pre-contact, pre-hunting diversity. He bought the spider monkeys at Panamanian markets and through private owners.

That the BCI troop wasn't "natural," per se, did have some bearing on Chrissy's project. How could it not? Normally, females emigrated from their troop in search of mates when they reached reproductive age. But on BCI, there were no other troops for females to join. Each of Chrissy's monkeys had known the others from birth, which probably affected their behavioral dynamics. Her male monkeys engaged in place-sniffing — a form of proceptivity in which they touched branches upon which a female recently sat, then sniffed their hands — and attacked females, just as in other populations, but no other spiders that Chrissy had read about engaged in clitoris holding, as hers did. She suspected that inbreeding — which lowered the sperm count — accounted for the troop's somewhat low birthrate. Just as an anthropologist could not study the mating behaviors of BCI residents and apply her findings to the patrons of Berkeley bars, Chrissy could not apply her Barro Colorado spider-monkey findings to other populations. Was she, then, on a fool's errand?

Some scientists believe that general principles — findings that can be applied to various systems — never work. Even among island systems, general principles are rare. Few spider-monkey populations occur in pristine forest, whether island or not. Even the monkeys in Peru's Manu National Park, often compared to BCI for its forest type and composition, experienced significant hunting pressure. Other spider-monkey habitats have been affected by agriculture or by fragmentation.

Chrissy acknowledged the constraints of her study, but she didn't let them worry her. "The situation is unnatural, yes. But it would be impossible to do this study elsewhere, because there's a lot more fission in other troops. And even though troops in other locations have smaller home ranges, they still spread out within them. It would be very difficult to follow your females."

Standing under the troop, Chrissy noted the time, the individuals she recognized — the cycling adult females Beri, Becka, Leona, Gracie, and Helena interested her most — and the fruit that was accelerating from monkey hands to the ground. At the moment, it was *Tetragastris panamensis.*

"It's a new fruit for them," Chrissy said, sounding excited.

"That's good?"

"Yeah. Maybe they'll stay in one spot for a while, until it runs out." She retrieved a fruit and picked off the husk. "Taste it," Chrissy said of the white aril, the flesh surrounding the seed. She tasted everything her monkeys ate. The *Tetragastris* was sweet.

As the season changed and the spider monkeys concentrated on new food sources, Chrissy would bring fruits to the lab for identification. But there her interest in forest ecology ended. Unlike most residents, Chrissy didn't share STRI's vision of BCI as a living laboratory for the exploration of forest dynamics and biodiversity. She had no STRI funding, and she was studying the sexual behavior, and that alone, of one organism.

The correlation of sexual behavior and hormones was Chrissy's particular puzzle piece, and she didn't waste a lot of time pondering its context: the architecture of the trees in which her monkeys lived, the phenology of their fruit, the symbionts and parasites associated with her troop. Established scientists often asked for her thoughts on the rate at which spider monkeys disperse fruit; she told them she had none.

A few minutes passed, and Chrissy started fumbling in her fanny pack with her right hand while holding her binoculars to her eyes with the left. She extracted a Ziploc. "Beri's gonna go," she said.

"How do you know?" I asked.

"Their anus gets all swollen-looking."

So it did. And soon the sample was falling toward the jumble of leaf litter at our feet. "Woo-hoo," Chrissy cried, jumping to her task. "You've got to get to it before the dung beetles arrive." I fell to, right behind Chrissy. It was exciting to be collecting fresh research material; it was exciting to be so intimate with nature.

Chrissy needed five grams of pure fecal matter, which meant that droppings larded with biggish seeds wouldn't do. She picked up several brown leaves and scraped them off inside the bag. She spotted another mass — "The mother lode!" — and knocked aside a dung beetle that was already in possession of a few milligrams. She couldn't afford to let more go. "Isn't this rewarding?" she said, squatting in the leaf litter.

"It is," I said. And I meant it.

Chrissy labeled the Ziploc, tucked it into her fanny pack, and moved on, smiling. There were more data to be bagged.

If scraping feces off leaves and noting how often a monkey touches its genitals seems an odd way to pass fifteen months, Chrissy didn't see it that way. The spider monkey's struggle to survive is a struggle to pass on its genes. And so a full understanding of the creature's reproductive biology — its sexual behaviors, breeding season, gestation length, and so on — is essential to knowing its ecology. And crucial for its conservation.

Spider monkeys, a valuable source of protein, experience severe hunting pressure. This, combined with their slow breeding rate (their interbirth interval is up to three years long) and their dependence on mature, undisturbed forest, has placed *Ateles* on the Convention on International Trade in Endangered Species list of endangered mammals throughout their range — from southern Mexico to the Amazon.

Naturally, Chrissy invoked conservation in her proposal to the National Science Foundation, which rewarded her with an $11,966 grant. Chrissy also won $7,000 from the Leakey Foundation, which funds the study of human origins. It is by no means easy for doctoral candidates to secure funding for their field research, but Chrissy had an edge: the general public has more interest in animals that are closely related to humans than it has interest in, say, arthropods. We fund the study of monkeys and apes because monkeys and apes are us.

"Spider monkeys developed social behavior convergent with

chimps, and chimps are humans' closest relatives," Chrissy said. (Convergence refers to traits developed in geographically or even temporally isolated animals — in this instance, chimps in Africa and spider monkeys in the Neotropics — not through a continuous evolutionary line but independently, in response to similar environmental challenges.) Shining light on the important determinants of social organization in spider monkeys, then, shines light on our own evolutionary history — an inadvertent intellectual bonus.

In her grant proposal, Chrissy had predicted that females who were periovulatory — meaning in the part of the cycle just before, during, or after ovulation — would be seen more often in subgroups containing males than at other times of the reproductive cycle; that females would have more copulations during this time; that the forest's typically diminished fruit supply between November and February would result in fewer conceptions; and that males would monitor females' reproductive condition via olfactory or behavioral cues.

That's where the clitoris came in. Chrissy hypothesized that in the absence of visual signs of estrus the female spider monkeys might be depositing scent cues with the clitoris. Over and over again she'd seen males place-sniff.

For comparison, a zookeeper at the San Diego Zoo was collecting urine and fecal samples from three female spider monkeys. This researcher had a far easier time of it than did Chrissy, as the captive monkeys, who lived with a vasectomized male, had been trained to urinate and defecate in a holding cage at the same time each morning. The fact that Chrissy couldn't get samples at the same time each day didn't bother her too much. The reproductive hormones cycled over a monthly period, she explained, and the daily variations shouldn't be great.

Tonight, Chrissy's fecal samples would go into the freezer on the second floor of the lab, and in December, when she concluded her field period, she'd cart a coolerful of frozen monkey dung through Tocumen Airport in Panama City and into a lab at the University of California, Davis. "The people in customs take one look in the box and they just send you right through," she said, waving her hand.

Unlike other doctoral candidates on the island, Chrissy wasn't getting daily feedback on her study, so she couldn't refine her techniques or hone her questions. She wasn't running an experiment, and the data she was collecting wouldn't mean a thing until she began to analyze hormonal metabolites and correlate them with behavior, eight months hence. At that point, the payoff could be significant. Chrissy could discover something unusual about spider-monkey reproductive biology. She could find that a particular food plant affected the females' cycling. Or that there was a staggeringly weird difference between zoo and wild animals. Such information would be a boon for conservation, and perhaps even for knowledge of human cycling and sexual behavior.

For now, though, Chrissy worked only on faith. If she was making a mistake in collecting the samples, she wouldn't know it until she analyzed them. Meanwhile, she was patient and confident that her samples would ultimately tell a compelling story.

We followed the troop's vocalizations — a whinnying sound, a long call, a Yoda-like trill, chatter — and its movements in the trees all afternoon. At times, the route was tortuous. The monkeys crossed ravines with a few graceful leaps between treetops; the hapless bipedal scientist had to maneuver down one sharp slope and then navigate back up another, all the while keeping an eye on the primate leaders.

Sometimes the troop moved so quickly I wondered if it was running away from us. Chrissy assured me, during a pause, that it was not. "They've been habituated to humans for years," she said. Was this a good thing?

When Carpenter arrived on Barro Colorado to study howler monkeys, he realized that his presence might distort the behavior of his study subjects, a phenomenon known to physicists as the Heisenberg Principle. To minimize the potential for perturbation, he wrote his field notes immediately but briefly, elaborating on them later. He constructed blinds in the trees and on the ground, then observed his quarry through field glasses. He hid behind banks, rocks, bushes, and tree trunks. Eventually he gave up on all this subterfuge. He found that after he stayed near the monkeys for long periods, daily, for a month, they reacted minimally to his presence.

In open terrain, wildlife observers may get away with blinds and ambushes. But in a dense forest like BCI's, where it's difficult to sneak around and spy on one's quarry for very long, habituation is mandatory. Chrissy told me, "If the monkeys aren't habituated to some degree, you can't study them. They'll just run away." She shrugged. "The Japanese study monkeys by setting up feeding stations." From the curl of her lip, it was obvious what she thought of that.

When the monkeys stopped to rest, we did too. "I think I'm developing ischium collosities," Chrissy said, stretching forward and moaning. She was referring to the calluses some monkey species develop on their behinds, which make branch sitting more comfortable.

We talked about food — a major preoccupation of BCI residents. Chrissy had some strong opinions on how French fries, scalloped potatoes, roasted potatoes, and chips ought to be prepared. We killed twenty minutes speculating on tonight's menu, during which her watch alarm rang twice.

It was time for a "scan." She wrote down what everyone in the subgroup she could see was doing — traveling, resting, feeding, defecating, grooming — and who was doing what to whom and for how long. She was particularly interested in behaviors sexual: place-sniffing, fondling, rubbing.

While Chrissy scribbled, I looked up. Usually I saw nothing, maybe a monkey moving from one tree to another. But in that movement Chrissy saw a male joining a subgroup that contained females, who may or may not have been receptive. She wouldn't know until she did the lab work a year later whether the female was periovulatory or not. The suspense made a useful blind, actually. Chrissy couldn't bias her observations if she didn't know what any of the behaviors were linked to. All she could do was note them.

In general, female monkey faces are not very distinctive, less so than males' faces. That made her first several months on the island difficult, Chrissy told me. "I'd come in so depressed because I couldn't tell the difference between any of the females. I wasn't getting any data." Then something clicked: the faces, and the differently shaped clitorises, began to emerge from the fur like a photograph in a darkroom tray. "It made all the difference," Chrissy said. "I started to have a whole lot more fun out here."

Instead of numbering her study organisms, as most scientists do, she gave names to the collections of eyebrows, cheek hairs, and clitorises. She named each monkey after a friend or family member, except Leona, who was the namesake of Chrissy's favorite Berkeley bar. Her study was now in forward gear. She began to see the monkeys' personalities and relationships within the troop, and her work became enjoyable.

Now, after hundreds of hours of observation, Chrissy read the spider monkeys fluently, without effort. She heard a chortling sound in the distance and she knew it was "just juvies, grappling." A monkey moved and she knew it was scratching, sniffing, preparing to urinate, signaling a juvenile. Every behavior meant something — perhaps it was trivial, perhaps it was heading toward statistical significance.

By late afternoon Chrissy had collected fecal samples from all five females, but we stayed with the troop until it settled down for the night, at around six. In other New World forests, spider monkeys do not sleep close to each other. In western Amazonia's Manu, for example, troops split into several subgroups. It has been theorized that a fissioned troop is less attractive to nocturnal scent-hunting predators. But on BCI, the spider monkeys' nocturnal scent-hunting predators — the jaguar and the ocelot — were long ago poached nearly from existence. And so the troop usually slept together, in one or two large trees.

Stowing her notebook for the day, Chrissy yelled "See ya!" to the silhouettes in the trees, and we headed toward the lab for dinner. In twelve short hours, she'd be back.

5

The Ingenious Habit

BRET WEINSTEIN lived in Kodak House, one of the wooden dorms in the old lab clearing, but his room was so far from the center of activity that his lab had become his de facto home. It was difficult to imagine anyone relaxing there, let alone working. Every surface in the small room was covered by electrical cables, cameras, batteries, coupling devices, hand tools, video cassettes, dirty laundry, clean laundry, shoes, towels, tripods, Ziplocs, journal articles, wine bottles, and urgent notes from other residents — "Bret, are you alive?"

I got a good look at this chaos the first time I hunted Bret down. I wanted to hear about his bats, which fashioned tents from leaves. Bret told me he was busy, though he didn't appear to be doing anything in particular. Clearly, he wasn't keen on talking to me.

A few days passed, and then Bret visited my lab. Leaning against the soapstone counter, he said, without smile or preamble, "Socially, it would help your cause on this island if you played in our Ultimate Frisbee game."

For years, BCI residents had played Frisbee on Friday and Sunday afternoons in Gamboa, the canal town sixteen kilometers east of BCI from which the island's launch embarked. It was an all-scientist game, co-ed and noncompetitive, though a few gung-ho males didn't suffer amateurism gladly. Bret was passionate about the game. When he'd arrived on BCI in March 1997, he'd brought, in addition to ninety-odd kilos of high-tech radiotracking and infrared photography equipment, his own Frisbee.

Bret's invitation to play surprised me, but if this was my only opening to the prickly bat researcher, I knew I'd better take it. "What time does the boat leave?" I said.

A day later, Bret again knocked at my door. "Do you want to see a bat?" he asked.

In the hallway, Heather Heying, Bret's girlfriend, gently pinned an *Artibeus watsoni* between her thumb and a small burlap bag. It was my first look at a tent-making bat. Bret pointed out the upright folds of skin on its nose, which are believed to help it to echolocate and, perhaps, smell. It had tiny white teeth. He spread the *watsoni*'s wings, about nineteen centimeters across, to show me the thinness of the membranes, the fine hand bones that acted as struts. The bat's order, Chiroptera, he told me, is Latin for hand wing. Natural selection had webbed and elongated the bat's hand for flight. This little bat weighed no more than ten grams.

Bret and Heather had captured the *watsoni* with a mist net on Donato Trail and were about to glue a transmitter to its back. "Neat," I said. I desperately wanted to witness this operation but didn't dare invite myself. I went back to work, wondering what had spurred this turn of events. Bret hadn't even seen me throw a Frisbee yet.

The next afternoon Bret returned, this time with a radio receiver slung around his neck. "Do you want to see where the bat's roosting? We just picked up the *watsoni*'s signal."

A hundred meters up Donato Trail, he switched on the receiver and began arcing an antenna in a ninety-degree sweep. The beeps led us off-trail, toward Lutz Ravine. In his rubber boots Bret walked with a slurp, troll-like, over the forest floor. He took another bearing and we turned left, contouring past a leaf-cutter ant colony. He took one more bearing.

"There it is." Bret pointed to a philodendron five meters

upslope. I didn't know what he was looking at. "See how the leaf tip is folded over?" He rifled through his backpack and pulled out a plastic saucer. "Walk up to it very quietly and look underneath."

As promised, there hung the *watsoni*, all folded up with a five-centimeter wire hanging from its back. Bret put the saucer under the roost, to collect the remnants of any fruit the bat fed on tonight.

Not five meters away, Bret spotted another philodendron that seemed to be in a state of modification. Turning the leaf over, he said, "See the roosting scars — those tiny little toe marks?" They were the size of pinpricks. "See the damaged veins?" The cuts didn't go all the way through because the leaf had to remain strong enough to protect the bats from rain. Charles Dominique, a French researcher, suggested that bats make tents as protection from weather and predators. But Bret didn't buy this: some tents were constructed with holes in the top.

Thomas Kunz, a professor at Boston University, has hypothesized that bats make tents as a defendable resource — in other words, to attract females and provide a place for their young. But Bret didn't like Kunz's hypothesis either, which holds that it is bats' polygyny — their pursuit of multiple females — that leads to tent making.

"There's a leap of logic there," Bret said. "Because you can just as easily defend your roost for one female." And while the tent makers *Euroderma bilobatum* and *Ectophila alba* are polygynous, *Artibeus phaeotis* and *Artibeus watsoni* looked, Bret told me, much closer to monogamous. He had never seen males of these species roosting with more than one female, though he couldn't be sure it was always the same female.

Walking back to the lab, ruminating on bats and on Bret's change of attitude toward me, I decided that he was like a wild animal, curious about me but easily scared away. And like an animal tasting an unfamiliar food, Bret was, for now, only nibbling, the better to avoid being seriously poisoned. I think I felt the same way about him.

Bret had been on BCI for just over a year. A Ph.D. student at the University of Michigan, he was writing his thesis on the evolution of tent making in bats. The behavior — biting and bending leaves into shapes that concealed and protected their occupants — had separately evolved at least three times in the entire life history of bats, and each cluster of tent-building species is far apart from the others on the bat phylogeny, or family tree.

"All tent makers are small, and all eat canopy fruit," Bret told me. We were sitting amid the rubble of his lab. His bare feet were nestled in a pile of dirty clothes. "It's interesting when evolution repeats itself with some pattern."

Unlike an experimental biologist, who devises experiments and predicts their outcome, Bret was working backward — retrodicting the process that evolution had wrought, trying to figure out *why*. Why tents? Why in the Neotropics? And why now?

Bret theorized that the bats construct tents near food sources in order to minimize their commuting costs. "Flight is very expensive for small bats," he continued. "The price to transport one gram of bat goes dramatically up as its size goes down. And these guys weigh only about ten grams. So I think they're making temporary roosts to minimize their commuting expenses between room and board."

The nature of fruit trees, he added, is another good reason for small fruit eaters to construct tents. "The resource is patchy — it's here and there in the forest. But 'here' and 'there' change as trees come into and out of fruit. If these animals used a permanent roost, they'd have to fly to the fruit-rich areas of the forest. By making a temporary shelter near fruit-laden trees, the bat gets many of the benefits of a permanent roost without the tremendous cost of flying back and forth." By pitching its tent fairly near a fruiting tree, then, the bat was simply camping near the pantry.

"The first tent I found on BCI was within thirty-five meters of a *Spondius mombin*," Bret said, naming a tree that produces orange fruits the size of kumquats. "It was within sight."

The bat could have made a tent even closer, but as fruit trees attract frugivores, they also attract animals that prey on frugivores. The distance a bat has to fly increases linearly as it roosts farther from a fruit tree, but the distance that a predator must search to find that bat increases by the square of that distance. The tent itself reduces predation — from snakes, opossums, coatimundi — by concealing the animal within. "Tent making is a more derived behavior than just roosting under an unmodified leaf," Bret said.

Still, tent making is restricted to only three branches of the bat's evolutionary tree. Why? Possibly, Bret thought, because tent construction and tent roosting have costs, and only in small canopy fruitbats are the savings great enough to justify the expense and hazards that come with the tent habit.

Large bats fly relatively cheaply, so reducing commuting costs is less of an issue for them. Other bats seek stationary resources or resources that require a fresh nightly search. Small fruit eaters fall in between. They need to find fruiting trees, but once they do, they eliminate weeks' worth of commuting costs by setting up camp. "It's a tradeoff," Bret said. Tents take hours and hours, over several nights, to complete. That's time away from food gathering, from mate seeking, and from, perhaps, mate guarding. "What I'd like to see," Bret said, "is the relationship between the amount of time spent in and around one tent and the amount of time that a tree is fruiting be very close to 1."

That's how Bret talked, like a scientist. And although he hadn't officially been inducted into that guild, he had thought of himself as a scientist for some time. He'd grown up in Los Angeles, stalking rocks, gems, and fossils in the mountains with his grandfather, a chemist. He took machines apart and he put them back together. He turned questions around in his head until he got them just right — shaped in such a way as to invite satisfactory answers.

His primary interest had always been animals. "I'd watch animals do things, and it made me think," Bret said. He wiggled a cable in his hands as he spoke. "As a kid, I once left cans of grape juice on a balcony overnight and discovered in the morning that raccoons had opened all the flip tops. I remember marveling at the fact that they had recognized cans as a food source. And how in hell did they open the tops? I always found it very seductive to figure out why organisms do what they do. Once my grandparents told me about natural selection, my fate was sealed."

At the University of Michigan, Bret became fascinated with bats after reading, in Richard Dawkins's *The Blind Watchmaker,* about the evolution of echolocation and its remarkable parallels with the invention of radar. He attended a bat lecture, saw a drawing of a bat tent, and decided this curious trait was something he wanted to study.

But with whom would he study it? Bret was concerned about becoming narrowly defined as an animal behaviorist. In a world that valued quantitative rigor, the subject was perceived by many as "soft." And so instead of joining the Organismal Biology subdepartment, Bret joined Ecology, a reasonable place to study bats, he figured: because they move around a lot of seeds, they affect the shape of the forest on a readily observable time scale.

It didn't take long for Ecology to find Bret disruptive. "They must have thought I was dangerous," he said. "I was always challenging them to consider ecological questions in an evolutionary light. To my astonishment, they insisted that one could study ecology separately. Even worse, they insisted that I not ask questions about how history had affected their systems. It was pure nonsense."

Bret gave me an example of the conflict. "I was interested in the forces that control the number of species in a forest. Ecologists are also interested in that question, but they want to talk only about the forces that allow species to coexist once they already exist together. I wanted to understand the process that put the species together in the first place. They had no interest in that. It's a bit like becoming a pilot and training to specialize in landings."

Bret turned to Evolutionary Biology, where he was assigned an advisor who specialized in crickets. Seeking expert chiropteran advice, he phoned one of the nation's top bat men, Charles O. Handley, Jr., curator of mammals at the National Museum of Natural History. After hearing Bret's theory of limited commuting costs, Handley invited him to BCI, to work for a month as his assistant.

In his element, the tropical forest, Handley was inspirational. "Around every corner," Bret said, "he found something exciting." As the two walked the forest, looking at fig trees and fig-eating bats, Bret pointed out modified leaves. Handley was at first skeptical that the leaves were in fact bat tents. Then Bret pointed out tents with architects in residence. Handley was subdued, "the way he is when he's intrigued," Bret said.

By the end of the field period, Handley realized that Bret had something to contribute. He suggested that Bret work on tent makers with Elisabeth Kalko, a scientist at Germany's University of Tübingen and his heir apparent on BCI.

The invitation thrilled Bret, not least because the graduate committee at his own university did not consider him ready to begin his Ph.D. fieldwork. It's a testament to the strength of Bret's nimble intellect and his powers of persuasion that Handley and Kalko — two highly regarded scientists with zero connection to the University of Michigan — came to sit on Bret's thesis committee.

BCI has always been very big on bats. The island is home to seventy species, and the initial discovery of tent-making bats in the New World was made in BCI's backyard, by the Harvard herpetologist and station founder Thomas Barbour.

In 1931, while visiting the nearby canal town of Pedro Miguel, Barbour noted groups of between nineteen and fifty-six *Uroderma bilobatum* roosting in the leaves of *Prichardia pacifica*, an introduced palm. The leaves, he wrote, seemed to have been modified by the bats themselves. Was it for maximum concealment? he asked. Was it for shade?

In his 1932 account of the finding, in the *Quarterly Review of Biology*, Barbour noted that bats aren't given to nest building, homemaking, or similar domestic activities. "By nipping the ridges of the plications on the under side," Barbour wrote, "the leaf is weakened and as the bitten spots are skillfully and serially distributed the leaf finally is sufficiently weakened so that the distal portion droops sharply downward." He deemed this tent making "a most ingenious habit."

The following year, Barbour removed all the modified leaves from eleven *Prichardia* in Pedro Miguel, then returned the next day and found a large frond, upright fourteen hours earlier, "completely nipped across its entire breadth, the distal portion fallen to form a perfect shelter and with eight bats in a cluster under the fold." Several other leaves were also newly modified.

Scientific proof that bats, and not some other creature, were building the tents called for observation, but Barbour's flashlight kept frightening the animals away. Constrained by the absence of nightscopes, ultrasonic microphones, infrared cameras, and super-fast lenses, Barbour's curiosity at this point rested.

Bat studies picked up again on BCI in the 1970s, when Charles Handley arrived to study BCI's fig-eaters. A formidable naturalist and tireless worker, Handley tagged and mapped every fig tree on BCI, thousands of them. Making periodic fruit censuses, he'd walk every island trail within two days, getting lifts from trail ends on boats. He was well into his seventies when Bret began working with him, and he still spent eight months a year in the field.

Over the years Handley and his colleagues had captured, tagged, and recaptured more than fifty thousand bats. This database, the world's largest on Neotropical chiropterans, lures a continuous stream of researchers to BCI. Kalko continues to expand that database, looking at one bat "guild" — a group of species with similar ecology — of seven species. Kalko's guild eats figs. Through bats, she is asking the classic questions of evolutionary biology: How do so many species coexist, since they are so similar, and why are bats so diverse?

Most of Kalko's bats belong to the family Phyllostomidae, which is endemic to the Neotropics — it occurs only here. Phyllostomidae means "leaf-nosed" and refers to the soft, spear-shaped folds of skin on the bat's face. By adding more surface area that sounds can bounce off, the fleshy facial feature — which looks radically different from species to species and makes all of them look pretty weird — is believed to enhance echolocation.

Because leaf-nosed bats carry fruits, katydids, or other objects in their mouths, they must emit their echolocation calls through their noses. The leaf is thought to focus or aim these signals. Like all New World bats, leaf-nosed bats use echolocation, in conjunction with a keen sense of smell, to maneuver in the forest and find food.

Looking at the role of echolocation in detecting, classifying, and localizing food, and its association with habitat type and foraging strategy, Kalko has discovered that bat sounds aren't as simple as had been assumed. Bats use frequency modulation, and they make "short" calls and social calls — both sonic and ultrasonic — to attract mates and mark territory. Kalko suspects they use short calls in other situations, whose context is yet unknown. Many species, Kalko found, change their calls radically when they switch from cluttered forest to a gap, from forest edge to open space.

Using the most advanced equipment, she is collecting data, for the first time, on the home ranges of her bats, their feeding habits and areas, their foraging strategies and the relationship of one species to another. She is also learning to recognize species-specific calls, a task much complicated by their high variability.

Phyllostomidae have radiated spectacularly throughout the New World: the family contains 143 species, who make their living off pollen, nectar, fruit juice, fruit, insects, frogs, lizards, fish, or blood. Here on BCI, a close relative of the phyllostomids, *Noctilio leporinus*, flies low over the cove at night. Known as the fishing bat, it rakes its long-clawed feet through the water, where sonar echoes reveal ripples caused by the surfacing of fish. Sitting on the raft at sundown, you could smell the bats' fishy odor as they swooped past.

BCI also has plenty of *Trachops cirrhosus*, which belong to the subfamily Phyllostominae and eat male frogs, homing in on the calls they make to attract females. The bats have tubercules on their chin that may act as chemical sensors when they touch a frog's poisonous skin. *Tonatia silvicola* makes its home in vacant arboreal termite nests, where it consumes katydids, moths, beetles, cockroaches, and other large insects. The nest offers the bat protection from its enemies: boas, spectacled owls, bat falcons.

Centurio senex, the "old man of a hundred," lacks a leaf nose but makes up for it with a deeply wrinkled face. When it goes to sleep, it pulls up a fold of skin from under its chin to cover its face. This sleeping veil has two translucent patches that cover its eyes like goggles. *Phylloderma stenops*, a scruffy-looking, spear-nosed bat, has been recorded feeding on the larvae and pupae of wasp nests.

Ectophylla alba bite heliconia leaves into a cozy tent where up to a dozen or more of these tiny, all-white bats roost. *Vampressa prusilla*, the smallest tent maker on BCI, is so skittish that it has so far evaded study. In Bret's experience, the bat is phobic about the sound made by flagging tape: it bolts every time he unspools his roll to mark their location.

After Heather went back to Michigan to pursue her own research, Bret needed a new triangulating partner. Late one afternoon he slipped a transmitter into his pocket, shoved a receiver and an antenna into my arms, and yelled "Find me!" as he dashed into the woods. It was a good game, a little like Marco Polo. But ten frustrating minutes of awkward advance and retreat passed before I could pinpoint Bret, hunched expectantly behind a fig tree. He offered some remedial coaching, and on my second attempt I cut my search time in half.

A few nights earlier, Bret had netted an *Artibeus phaeotis* — a small, frugivorous bat common to BCI, a cousin to the species he'd shown me in the lab. Bret trimmed some of the soft brown fur from its back and glued a one-gram transmitter to the spot. The transmitter was the size of a raisin, with the nine-centimeter antenna attached.

As the bat went about its business over the next several nights — finding fruit and carrying it back to a feeding roost to eat — Bret and I would triangulate its location. Working with compasses from two known points along Donato Trail, we would extrapolate a third — the bat's.

Now, on our third night triangulating together, the receiver at my side pulsed concisely and constantly: "Beep. Beep. Beep. Beep. Beep."

"Roosting," Bret's voice crackled over the walkie-talkie.

A minute passed, then "Beepbeepbeepbeepbeep."

"Flying," Bret said. The pulse speeded up when the long tail of the transmitter moved from a vertical to a horizontal position. I flicked on my headlamp and rose from my overturned bucket. Atop the bucket was a pink life vest, for comfort, and under that, some plastic sheeting, for chigger prophylaxis. I already had two dozen bites.

As on previous nights of triangulating, the forest air felt close and damp. The plastic strap of my receiver was wet with sweat. The butt end of the antenna, which was about a meter and a half long, with four horizontal branches, dug into my hip bone.

"Are you ready?" Bret asked. Neither of us announced our compass point before the other was ready, to avoid influencing the readings. I twisted the dial of the compass, duct-taped to the receiver handle.

"I've got 282," I said into the walkie-talkie.

"Gotcha," Bret said. Three hundred meters away, he wrote his coordinate and mine in a logbook. Later he'd plug the numbers into a computer program that plots the triangulation points on a map of the island.

Now the beeps slowed down. "Roosting," Bret said. On our own, we slowly turned our antennas in an arc, listening to the strength of the beeps and watching a needle bounce on the receiver's dial. In the ninety-degree sweep where the signal came in strongest, we gradually turned down our filters, a scintilla at a time, until we could narrow the arc to a fixed point.

Listening to the receiver was tedious and frustrating: the signal came and went, even when the bat wasn't moving. Small transmitters send weak signals, and the island's geography and vegetation can distort the sound waves. Sometimes we isolated the exact bearing quickly, but it often took us a minute or two. An exact bearing was at times impossible. When the bat flew, the pressure was the highest: without a nearly simultaneous read, the triangulation would be meaningless.

I got my bearing and turned off my headlamp. I was saving battery juice and trying to minimize the number of insects flying into my nose and eyes.

"You ready?" Bret asked over the walkie-talkie.

"Yes."

"Tell me it isn't 340." That was what I'd been getting quite frequently.

"No. It's 274."

"Good God, it must be Christmas."

I was on tenterhooks. "That's good for you?"

"Yeah, that works." His number had also changed, which meant the bat was feeding in more than one roost.

Our bat had been flying for up to seven minutes at a stretch, almost twice as long as other *Artibeus phaeotis* that Bret had tracked. The long flights suggested to Bret that the bat was having a hard time finding ripe fruit, and so it kept circling and circling the known resource patches near its roost.

I pressed the talk button. "Why would the bat be searching for so long if fruit is abundant now?" It was the beginning of the rainy season, and the forest seemed full of food.

"I'm not sure that the fruit these bats prefer *is* abundant," Bret said. "They eat just a subset of the forest. Over and again I've seen it bringing back to its feeding roost fruits it doesn't ordinarily eat, in addition to fruits that aren't ripe."

Unlike Chrissy, who kept the focus of her spider-monkey study narrow, Bret was eager to examine possible puzzle pieces surrounding the tent-making habit. He considered his bat's relationship with its environment, the phenology of the fruit it ate, and its competitors and symbionts.

In the afternoon, when he woke up, it was Bret's habit to hunt down tents made in palms, philodendron, heliconia, and aroids. The leaves were large and stiff, bitten into boat shapes, cones, and A-frames. Under roosts where *phaeotis* had been feeding, he scouted for shriveled pulp among the leaf litter. The bats chewed off small bits of the fruit's flesh, sucked out the energy-rich juice, and dropped the rest — which made deconstruction of the frugivore's diet refreshingly simple.

The question tonight, for Bret, was how far his bats would fly to find the food they did prefer. But we'd spent four hours triangulating, and the bat didn't seem to be doing anything new or interesting: it roosted in the same place for twenty minutes and made its seven-minute flights in the same general area. Still, Bret needed a full night's data. The bat could change its behavior at four A.M., and if Bret wasn't there to document it, the evening's exercise would be meaningless.

We plotted on, each of us alone in our own private darkness, inured to the ceaseless beeping of the receiver. It was hot, and it was buggy. But sitting in the inky blackness of the forest was not without reward. One night I turned on my light just as a meter-long snake, thin as a pencil, green as a vine, glided past the tips of my boots on a search-and-destroy mission for frog. Another night was punctuated by the high whistles of kinkajous, up in the canopy. Kinkajous (*Potos flavus*) belong to the family Procyonidae: nocturnal and omnivorous, they are cousin to the raccoon. They look slightly monkeyish, with rounded ears set on the side of their face. When I turned my headlamp to the trees, the light caught some yellow eyeshine but never the actual beast.

It was said that night monkeys, or douroucouli, also inhabited BCI, but no one had seen this animal for years. The world's only nocturnal monkey, it weighed just under a kilogram (about two pounds) and had huge eyes and a rounded head. Martin Moynihan believed they communicated with large physical gestures, rather than facial expressions, as an adaptation to life in the dark. A researcher who boarded in Gamboa told me that night monkeys lived in a hollow tree just down the road from him. By day we stalked the tree, smack in the middle of a neighbor's yard, but saw no evidence of douroucouli.

Bret's voice sang out through the dark. "Do you have those little white flies up there? Taking small bites of your flesh?"

"Yup," I answered, examining two dots of blood on my arm. "When you turn on your light do you get little cockroaches crashing into your face?" I could never forget that BCI had a hundred species of cockroaches, my least favorite insect.

"Yup."

And so the hours passed.

It was during these long nights of tracking that I got to know Bret. I learned, in fragments over the walkie-talkie, that he did poorly in school; that he was a fidgety kid and he needed always to be holding something. Often it was a part he'd extracted from a radio, camera, or phonograph. At first he got in trouble for taking things apart. Then he started to think about how machines worked and tried putting them back together. He got better at it.

"For as long as I could remember, what seemed cool was the power that came with recognizing good and bad questions and learning how to answer them," Bret said. And so he knew, from early on, that he wanted to be a scientist.

He attended a private high school, got bad grades, and was asked to leave. He started over at another private school. In tenth grade he was diagnosed with dyslexia. Still, he was learning. "People knew I was smart," Bret said. "They'd ask for my notes, thinking they must be good. I'd say, 'I don't have *any* notes, man.' " At the start of every term he vowed to do better, but he'd realize, within a few weeks, that it wasn't going to work. And he'd give up.

Bret passed the time staring at the ceiling, thinking. He kept taking things apart. He challenged authority — in the pro forma way of adolescents, but also as a young man in love with logic and falsification, the verities of science. Falsification states that a hypothesis is acceptable as long as it has withstood all attempts to prove it false, either by experiment or by contrary evidence.

Bret studied himself — his sleep-wake cycles, his reaction to nicotine and to hallucinogens. He and Heather went east to college. She hated Smith; he dropped out of the University of Pennsylvania, harassed by frat boys. Together they drove west, settling in Berkeley. Bret repaired bicycles; he considered college "just another entitlement, a way to reward people who are going to be rewarded anyway." Eventually he earned his bachelor's in biology at Santa Cruz.

"I had problems with the system, I always did," Bret said. "But Heather didn't: she did very well in school, so I figured it couldn't be all bad. I just had to find a way to rail against only the broken aspects of it, not all of it."

Together they enrolled at the University of Michigan: Heather started a Ph.D. on the evolution of mantelline frogs, the Old World analog to the New World's poison-dart frogs; Bret began to dismantle tent-making bats.

Another half-hour passed: flying, roosting, flying, roosting. We were getting some nice data now. Suddenly a troop of howler monkeys let loose with an ear-splitting scream. "Get ready for rain," Bret said, tersely. Howlers were never wrong about impending storms. Soon, every troop within earshot was bellowing away. The most ferocious sounds came from males, who have a throat sac that acts as a resonator. The howls carry for nearly a mile through the forest and act as a territorial warning: troops that announce their turf can avoid battling over it.

Just as I pulled up my hood, the sky opened. At first the rain only splashed onto the sheltering leaves overhead, but as it intensified, the canopy seemed to wither in defeat. Water pelted my back, forming rivulets that soaked my trousers and ran into my rubber boots. A sudden bolt of lightning illuminated everything around me, including the metal antenna that lay against my knee. Whoa, I said to myself, jumping up to lay it on the ground.

"You okay?" Bret asked.

"Yes," I answered. "You?"

"I'm all right, but there's a little bit more lightning out here than I'd like." His voice came to me between crackles of lightning. "What do you think?"

I was thinking this was almost miserable. I knew the bat wouldn't fly through the heavy rain, and that the storm would eventually pass. Mostly, I didn't want to be the first to cry uncle. "Let's wait," I said.

"Okay," Bret said. He sounded resigned.

About a month earlier, when it looked as if Bret would have nothing to do with visiting writers, Bert Leigh had said to me, "That's too bad, because Bret is a whole lot smarter than most of the people here. But at least you've gotten yourself out of a lot of odious night work." When I had told Bret the story, he'd laughed.

Now, we made it through another fifteen minutes of downpour, with abundant lightning. And then the rain, devilishly, intensified. Small rivers coursed down the trail, depositing moraines of debris here and there. The raindrops felt as big as plug nickels. My raincoat no longer repelled. Through the cacophony I heard Bret's voice, a small sunny thing in the dark. "So, Elizabeth. Is this odious enough for you?"

Finally, it was. We packed up the equipment and hiked back to the lab.

After several nights of netting and tracking bats, I wondered if Bret envied biologists who got their doctorates studying captive animals, or who did their research sitting on a warm, dry lab stool, looking through high-powered microscopes at tissue that didn't run or fly away.

There was untold griping on BCI about molecular biologists because they got a lot of money and a lot of prestige, largely by staying indoors. But field biologists, when pressed, acknowledged that molecular biology was actually invigorating natural history, opening new avenues of inquiry and closing old paths of speculation.

Information from molecular analysis has made some current descriptions of animals, and their evolutionary and real-time relationships to other animals, obsolete. For example, molecular taxonomists now claim that hippopotamuses, once believed to be close relatives of pigs, are actually closer cousins to whales. Assumptions about sexual behavior have been turned on their heads. Until DNA analysis told us otherwise, it appeared that certain bird species mated for life.

In the stairwells of BCI's dorms roosted a dark gray, medium-sized bat called *Saccopteryx bilineata*. The males and females, which were insectivorous and roosted in harems of one male and up to eight females, had a faint pair of white lines running down their backs and, on their wings, a small pouch of skin. In the male, this sack contained a scent gland, from which the bat wafted pheromones toward females.

Gerald Heckel, of the University of Erlangen, near Nürnberg, was studying reproductive success in BCI's *Saccopteryx*, and his primary tool was the biopsy punch, which snipped a small circle of skin from a bat's wing. The damage was not severe: the hole, the size of a pencil eraser, would close long before Heckel wrote up his work.

By examining the DNA in the wing cells, Heckel could determine the relatedness of the bats he surveyed. He could find out if males that guarded harems of females were better at passing on their genes than were solitary males. A hundred years ago, or fifty or even twenty-five, a bat scientist couldn't have gotten this information. To determine whether the colony's males or females dispersed at maturity, he would have to capture, mark, and track individuals. Today's researcher can simply capture and snip; lab analysis would reveal a colony's genetic structure.

A colony in which there was little or no dispersal would be highly structured: that is, its members would have a lot of the same alleles — any of the different forms of a gene that occupy the same locus on a chromosome. Inbreeding would be high. Conversely, a colony in which both sexes dispersed would have low structuring.

Heckel was looking at mitochondrial DNA, which passes only through maternal genes: if females were dispersing, then he'd see their genes in numerous colonies. If females were staying home, he'd see an individual's mitochondrial structure only in her colony.

Heckel was also examining the degree of inbreeding among *Saccopteryx* colonies. The bats had small territories, and it appeared they didn't like to fly over water. Had the population on BCI become inbred since it was isolated as an island? To find out, he was comparing genetic diversity here with the genetic diversity of sac-wing bats in La Selva, a Costa Rican reserve that is surrounded by forest.

So far, Heckel's work had only raised more questions. He found that seventy percent of bat babies in a harem were not fathered by the resident male adult. He filmed the bats during mating season and saw no sex. Why would a male bother defending his territory against other males, in and outside of breeding season? Heckel saw no male parental care of bat babies, so why would females stay in the harem after they'd given birth? In a harem situation, males are usually bigger and stronger than females; they've been selected for size. But the female *Saccopteryx* around the BCI dorms were the larger, stronger animals. Why?

In one paper Heckel concluded that males *try* to increase their chances of mating by keeping a harem, but their ability to control females is not high. Basically, the females are off having sex in the woods. Heckel also found that males have a higher success rate if they defend a harem than if they're transient.

I mentioned Gerald's DNA work to Bret and asked him why a male bat would go to the trouble of defending a crowded roost if he wasn't passing his genes along. A look of consternation crossed Bret's face. Referring to the lab work, he said, "He's not doing it right. If an expensive behavior like that didn't facilitate passing on the male's genes, then males who didn't bother would have a great advantage and natural selection would eliminate it."

I was reminded of Mark Twain's remark: "Researchers have already cast much darkness on the subject, and if they continue their investigations we shall soon know nothing at all about it."

Could DNA help Bret figure out why small, fruit-eating bats make tents? Would it make his long nights of observation obsolete? Hardly. The tent-making trait wasn't attached to a single mutation, or perhaps even several genes. "Tent making might not even be genetic," Bret said.

Microsatellite analysis, an expensive and labor-intensive process, might tell Bret the most recent physical place in the bat phylogeny that tent making evolved, but it wouldn't reveal if this was the first instance. "Tent makers could have gone through a bottleneck, where all but a few individuals died out," Bret said. In this scenario, an ancient tent-making population may have experienced a drop in population, with a corresponding drop in genetic variability. As the population grew over millennia, the trait may have become reestablished.

Even if Bret could pinpoint when the ingenious habit first appeared, the knowledge wasn't all that useful to him: it still wouldn't tell him why the trait arose, because he wouldn't know what that progenitor bat's environment was like. Thirty million years ago, what did bats eat? Who were their competitors? The answers were so speculative that the question was nearly worthless.

Still, DNA analysis could be helpful to Bret. If he analyzed tissue from pairs of bats with offspring, he could, like Heckel, learn about paternity and the social structure of his organisms. Were they pair-bonded? For how long? If it was short-term only, were male bats frequently investing in tents that protected their predecessor's offspring?

For now, Bret was making do with the basic tools in the evolutionist's kit: comparative analysis, information on geographical distribution, and logic. And, of course, long, long hours of observation in the organism's natural habitat.

The prospect of all that time in the forest thrilled him.

6

Rat Patrol

I was going out a couple of times a week with Paul Trebe, who'd been the first person to offer me a job on BCI, and getting a lot of exercise. The mornings all started the same: Paul wolfed down his breakfast, loaded the gas can into his boat, and met me on the dock at seven with pillowcase in hand.

The case, at this hour, weighed 100 grams. By eight, Paul would be sweating so much that it weighed 110 grams. By the time he was finished with the third island of the day's census, the pillowcase would reach 130 grams. It was important to pay attention because if we didn't accurately weigh the spiny rats scrambling around inside, we could misinterpret their age. Which would, in turn, skew the demographic snapshot of the captured population's juveniles, subadults, and adults.

Today we were running the traplines on islands 14, 4, 12, and 9 — the worst group of his twelve. The islands were small, but two were so steep it took handholds on roots and vines to get us to the ridge and back down. The transects — parallel trails that ran from one end of the island to the other — were lightly etched and muddy; spiky vines grabbed our arms, roots tripped our feet. We were both flecked with dirt and sticky with banana slime. Every twenty meters along each transect, we emptied a trap and rebaited it with fruit.

This job was mostly about stamina — one island had ninety-nine traps — but it demanded, also, some finesse. At each station Paul bent over, placed the mud-streaked, blood-stained pillowcase over the business end of the trap, flipped open the mesh door, turned the trap 180 degrees, lifted it to his mouth and exhaled sharply onto the rat's hind end. The startled animal scrambled out of the trap and plopped into the pillowcase, which Paul clipped onto a pocket-size metric scale.

The first time I saw Paul put the cage to his face, I asked him if he'd read *1984*, the scene where Winston Smith's torturer straps a wire cage, with rat inside, to his face. "No," Paul said, smiling quizzically. "I never got around to reading that one."

I liked working with Paul, but he wasn't the lead scientist on this project. He was a field assistant to a University of Wisconsin ecology professor named Greg Adler, who'd set up this study years ago and now visited Panama two or three times a year. So, basically, I was playing field assistant to a field assistant. My job consisted mostly of writing things down, but it saved Paul at least an hour on his rounds, sometimes two.

Feeling carefully from the outside of the pillowcase for the rat's neck, Paul grabbed a fold of skin and flipped the case inside out to expose the animal's white-furred belly. While I wrote in his logbook, Paul read off a four-digit number, corresponding to the combination of rat toes that had been clipped near the second knuckle.

Toe clipping, Paul explained, is the only good way to differentiate among individuals who will be studied for many seasons. Metal bands would rust, ear tags fall off, bleach jobs grow out. The procedure, performed with cuticle nippers, was relatively painless, according to Paul, and it followed federal guidelines for animal care and use. After being snipped, the rats dashed off through the leaf litter. Paul had never seen an infected foot, and altered toes didn't seem to affect the rat's livelihood. So far, the spiny rat project, in seven-odd years, had left its indelible mark on more than seven thousand rodents.

Whenever he came upon a new, unmarked rat, Paul's lips tightened. He had no taste for blood, for deliberate injury.

"Some of them — the males — squeal a little, but there isn't much blood," he said. "Still, this is the worst part of the job." He wanted to teach me to do it, but it made me queasy, the sound of snapping bone. When I did eventually clip a rat, Paul said to me, "Good. Now if I go to hell for this at least I'll see you there."

After I entered the rat's ID number in the logbook, Paul read off the rat's other vital statistics. "Female. Nonperforated, small, closed. Complete tail." He was telling me about the rat's pubic symphysis, whether it had been opened through previous births or never opened, and whether she was pregnant. Because tails break off easily, presumably as an antipredator defense, some spiny rats had just a nubbin protruding from their rear end.

For male rats, I wrote down whether their testes had descended — S for scrotal — or not — A for abdominal. Then we dropped the rat back inside the pillowcase and weighed it. Babies came in at about 100 grams; the monster rats weighed more than 700.

Spiny rats — *Proechimys semispinosus* — are brownish, docile animals with white underbellies. They get their name from the streak of longer, stiffer hairs that run down their backs. Paul had never seen a spiny rat raise its hackles. Most of these rats were so accustomed to capture that they just relaxed in his hand until he returned them to the leaf litter. Sometimes they jumped right into another baited trap. "Do that after I'm gone and you're gonna have to wait till tomorrow to get out," Paul warned the recidivists.

Nocturnal creatures, spiny rats spend their days in cool, shallow depressions under roots, in hollow trees, or in burrows made by other animals. They have long, narrow heads, large eyes, slender ears, and whiskers that extend to their shoulders. The guidebook *Neotropical Rainforest Mammals* notes that *semispinosus* are "large, handsome rats." Tasty, too: their meat is said to be excellent, and they're often trapped for food and sold in markets.

Paul was a brown-haired, fairly docile recent graduate of the University of Wisconsin, Oshkosh. He was small on small talk, solid, and calm. At least, he was these things in the morning. The attitude of his short hair made a good indicator. Before the sun reached its zenith, it lay flat, in an exaggeratedly neat style. After the sun set, Paul's hair got rowdy, spiny even. His jokes took on color, his attitude grew bolder. As the beer flowed, Paul reverted to type: a red-blooded frat boy and former bartender from Wisconsin.

It was only fitting that Paul's other job on BCI was beer *jefe*. He was responsible for ordering forty cases a month, keeping the beer machine full, and returning empties to the Cerveceria Nacionál. The island also had rotating jefes for stocking the store, making change, and procuring videos. Of all these jobs, that of beer jefe was, naturally, the most prestigious.

In Wisconsin, Paul had been looking for work when his ecology professor, Greg Adler, asked him to run the Panamanian traplines. With the help of assistants, Adler had been censusing rats on twelve islands of between two and four hectares since 1991, when he also measured and mapped every fruiting tree greater than ten centimeters in diameter at breast height. The job also entailed taking a monthly fruit census to determine which trees and lianas were currently producing the staples of the spiny-rat diet. From the data that Paul e-mailed to Oshkosh almost daily, Adler would conjure a snapshot of rat population size, age structure, body mass, sex ratio, reproductive output, and survivability of offspring. From such data he would, over the years, make much hay.

Although his proximate study subjects differed, Adler was, ultimately, mining a vein of inquiry similar to that tapped by Robin Foster in the 1970s. Like Foster, Adler was interested in the relationship between animals and their food supply. To understand rat ecology, he had to do what Foster had done: identify the fruits that fell on his islands — using Thomas Croat's *Flora* and herbarium samples that Foster himself had gathered — and determine who was eating what, and when. Eventually, he'd discover not only that the forests on these tiny islands were shaping rat populations, but that the rats were themselves shaping the forests.

As Darwin found in the Galápagos, islands — with their limited immigration and emigration — make perfect laboratories for getting at universal truths. Scientists use them to study minimum and maximum critical populations, the effects of invasive species, and how plants and animals disperse and evolve.

Adler's Gatun Lake replicates were a perfect setup for playing all manner of intellectually satisfying games. By manipulating their populations — which averaged twenty rats per hectare — and making comparisons between islands, and between island and mainland populations, Adler could look at territoriality, at the impact of resource abundance on population density, at how rats change their reproductive rates in response to density, and at the effect of limited competition on population stability and reproductive effort. Far-reaching and long-running, Adler's project was a paradigm of ecological studies.

In his first five years in Panama, Adler made more than 15,500 captures. He established that island rats conform to what's known as the island syndrome: they grow larger than their mainland counterparts and their reproductive effort is lower. He established also that island rats live in higher densities than do mainland rats, as a result of their isolation (which keeps a lid on their dispersal) and the low intensity of predators and competitors — known in ecology as "density-depressing agents."

Adler turned out to be a slightly gonzo-looking man with seventies hair and a Pancho Villa mustache. When we first met, months into my stay, he was wearing a loose, floral-patterned shirt, shorts, and Tevas, one of which was held together by a segment of bandanna. Adler was passing through Panama on his way home from Venezuela, where he'd spent a few weeks looking at forested islands in Lago Guri, a reservoir dammed in 1986. The islands were rodent-heavy, and so they appealed to Adler. He thought they'd make an attractive site for comparative studies.

Growing up in West Virginia, Adler spent a fair amount of time setting traps for rats and mice. "I always liked them," he told me. "They're model organisms. They're good when you're looking for patterns, and you can extrapolate to other rodents and mammals, within limits."

Adler began studying rodents in earnest as a Ph.D. student at Boston University. For three and a half years he visited Nantucket once a month, censusing white-footed mice and meadow voles. He did a postdoc at Harvard, studying mice in Ipswich, Massachusetts.

With grants from the National Science Foundation and the National Geographic Society, he arrived in Panama in 1991. The project on BCI, he said, "is a more detailed, advanced version of my postdoc. Each project leads to another, from temperate to tropical and then back again."

By working brutal hours (for the first three years Adler employed no assistants) and doing experimental work at a time when ecological experiments were freshly in vogue, he rose quickly in the field of population biology. He published his results in the *Journal of Animal Ecology*, *Oecologia*, the *Journal of Zoology* (London), and other scientific periodicals. By the time the University of Wisconsin hired him, in 1993, no other scientist had studied a tropical rodent so thoroughly and for so long.

The big question for Adler has always been, What limits the population densities of tropical animals? The spiny rat was well positioned to provide answers. It was the most common rodent throughout much of lowland Central America and northwestern South America; its natural history was well known; and it was easily sampled. Adler knew that spiny rats are primarily frugivorous and that their population density and reproductive activity cycle in accord with yearly fruit production.

Among Adler's more important findings was that even in the season of fruit abundance, rats are food-limited — that is, the availability of fruit keeps their population in check. To ascertain this, he spent from May until October of 1992 placing rations of fruit in ten feeding stations per hectare, three times a month, on four islands. Four comparable islands served as controls. He placed the fruit in wire cages that admitted rats but not agoutis or pacas. After lengthy analysis of the data, Adler concluded that the per capita increase of known births within the experimental populations was nearly double that of the control populations.

"The first time I gave them food, the results surprised me," Adler said. "I didn't expect the numbers to go up. They went up because of increased reproduction, not better survival of the adults or because they were growing bigger or faster." In other words, adult females simply had larger litters; they neither lived longer nor reached maturity any quicker — two ways of boosting population size.

"In November, December, and January of 1997, and for those months in '98, I'm provisioning the rats again, but in a time of resource scarcity," he continued. "The rats don't do well then: they have more mites, their hair falls out, they lose weight." As the rainy season progressed, I would see these reddish mites, as well as a fungal growth, around the rims of the rats' ears.

"I didn't know, back when I started, that the project would be so long-term and what sort of questions I'd be asking today. Of course, these studies all hinge on the availability of funding. But there's a nearly endless supply of questions I could ask. I could work on genetics; I could look at the role of rats as reservoirs for infectious agents."

With funding from the World Health Organization and the

Colombian government, Adler was, in fact, looking at rodents and possums as vectors for *Leishmania*, the protozoan that causes the leishmaniasis disease. At Harvard, he was studying Lyme disease. The epidemiological research was gratifying, he said, but its flat-out utility had a bonus: it gave him something to chat about at cocktail parties.

"I used to talk to people about the role of rats in seed dispersal, but they'd just say, 'Well, what's in it for me? Why bother studying that?' Now I just tell them I study diseases in rats. Basically, though, I'm an ecologist. And I like rodents."

Letters and numbers filled the columns of Paul's logbook. I wrote, I emptied traps, I rebaited. Like automatons, we moved from one station to the next. Now and then, something anomalous in the trap surprised us: a *Bufo marinus*, a four-pound warty brown toad, or a brilliant yellow butterfly, interested in the fruit's nutrients. "Look at that," Paul would whisper, pointing to an olive-blue bird on an overhead branch. "See his thin tail? It's a motmot." Once Paul startled a huge iguana that catapulted off its branch and straight into his chest, almost knocking him down the slope he'd just climbed.

Most of the time, though, the work was mundane, and it was easy to get resentful of Adler, sitting in air-conditioned splendor in his office back home. He wasn't testing a hypothesis at this time, or running a manipulation, but he still needed Paul's data. Why?

Because Paul's numbers were background hum, the stage setting that allowed Adler to launch a new experiment at any moment. Even if he wasn't plugging Paul's numbers into mathematical equations today, the daily catch and the state of the islands' fruit production still had significance: it was baseline data from which Adler would, when running an experiment, draw comparisons.

Doctoral candidates like Chrissy and Bret relied on BCI's baseline plant and animal data, but their smaller-scope work, by necessity, added only snapshots of information about their study organisms to the island's trove. Adler's work, however, was a dramatic example of how a researcher used baseline data in order to create more complex data. Using the information assiduously collected on rats and fruit before starting the food supplementation experiment, Adler could compare treatment islands to control islands, and control islands during the experiment to the same islands before the experiment. He could even compare his islands to BCI.

From Joe Wright, the STRI scientist who runs the island's Environmental Sciences Program, Adler learned that fruit production on Barro Colorado, less than 152 meters from his experimental islands, was greater in 1992 — the year he first supplemented the rat diet — than in any year from 1984 through 1994. That meant that his islands might have experienced an even greater jump in rat population density had he conducted his experiment in a more "average" year.

Having more numbers at his fingertips meant that Adler had more comparisons at his fingertips — different, and perhaps better, ways to interpret his results. The numbers were like flecks in a kaleidoscope that Adler could turn, at his whim, to achieve novel perspectives on the same phenomena.

Without Paul, though, such richness of opportunity would be nearly impossible. After all, Adler had two classes per semester to teach in Oshkosh, data to analyze and publish, and grant proposals to write. In the university's view, professors who don't bring in grant money and burnish the institution's image with their publications are hardly worth keeping around.

Paul and I sat in the boat catching our breath. We'd jogged the last few transects — for the pleasure of feeling ourselves move at more than drudge pace, and to get it over with sooner. Paul picked from his palm a broken-off spine of an *Astrocaryum standleyanum*, or black palm, whose trunk was uniformly studded with these sinister slivers. Paul had warned me that *Astro-*

caryum cuts always get infected, because the spines are covered with an irritating chemical. *Astrocaryum* was an excellent provider for rats, dropping between 1,800 and 3,000 large fruits a season.

Taking a swig of water, Paul wrote in his logbook, "Traps left open, 11:35." We'd be back tomorrow, to do the whole hellish set of islands again, then again the next day to remove rats and close the traps until next month. Meanwhile, he'd open traps on another four islands. Paul sighed.

"Have you ever considered quitting?" I asked.

"In the beginning I thought about it, when I was hauling all that corn around the islands, another supplementation experiment. But I could never do that." He shook his head. "I'd be too embarrassed."

"Will you go on to grad school, to study tropical biology?"

"No," he said quietly. The fieldwork was too tedious for him, and he didn't like the climate. "But this is a good experience," he added, "and I'm glad I'm doing it."

We motored slowly away from the island, threading our way through the drowned tree stumps. Ordinarily the stumps were a meter or more under water and posed no navigational hazard. But the water was now at its lowest level since the lake was formed. Because of last year's El Niño, the island received just 1,700 millimeters of rain in 1997, instead of its usual 2,600, and this year's dry season was the driest on record.

In Laboratory Cove, Slothia, a 0.3-hectare islet, was now attached to *its* mainland, Barro Colorado. Grass had sprouted on the tiny isthmus, and river otters had been using it as a gangplank. Madden Lake, which supplied the water to operate the locks, was almost empty, and the Panama Canal Commission had recently begun siphoning Gatun Lake to take up the slack. Every boat transit drained 52 million gallons. Already, freighters were offloading excess fuel before they made their passage, anything to lighten their draw. There was talk the canal might have to close.

For residents of BCI, a low lake level meant that boat propellers nicked exposed stumps, and that mats of hydrilla, an aquatic weed, were more concentrated below the lake's surface. The strain on the boats was obvious: Ricardo Cajár and his team of mechanics were now fixing three transmissions a month. Paul told me that the lake was expected to drop at least another two meters by the end of April.

The rainy season would start at the end of that month, but it would bring no immediate help. Thanks to scientists who study rainfall patterns and such esoterica as the evapotranspiration rates of trees, the Panama Canal Commission knew that the forest, once the rains began, would take its cut of water first. The parched soil would absorb the first month of rainfall; there would be no runoff until the forest was satisfied. It was an object lesson in how basic research paid off: without a minute examination of how water moved through a tropical forest, a captain who planned to nip through the Isthmus of Panama, for example, might find himself unexpectedly circumnavigating the Southern Hemisphere.

Paul opened the throttle as he turned into the shipping channel. The wind had come up and the little boat pounded over the fresh chop. Paul slowed for nothing: it was lunchtime. Afterward, he'd play a quick computer game, change his T-shirt, and wearily head out to census fruit on another group of islands.

I'd help Paul again and again in the months to come, but only when I felt like it. Paul had no such choice. Back in Wisconsin, Greg Adler was waiting for data.

Fifty years ago, Frank Chapman asked what mechanism controls vertebrate populations on BCI. Why were there so many coatimundi here? he wondered. Would they eventually saturate the island, and would their population someday crash?

In some forests, it's been theorized, such large predators as pumas, jaguars, and harpy eagles keep a cap on populations of smaller vertebrates. But BCI's large predators, even in Chapman's time, had been hunted nearly to extinction. The fruit famine of 1970, in which so many fruit eaters died, suggested another mechanism to researchers. By controlling their pollinators and dispersers with shortages of fruit, nectar, and new leaves, BCI's plants control BCI's vertebrates.

It has long been known that bat-pollinated trees secrete only tiny amounts of nectar throughout the night so that their visiting bats take a sip and then go away, to other receptive flowers of the same species. Nicholas Smythe, a grad student working on BCI in the late 1960s, showed that fruiting trees hold similar sway over agoutis.

These large rodents are the principal disperser of seeds too large to be swallowed by most animals on BCI. The seeds, clustered around the base of a parent tree, risk heavy mortality by predatory herbivores and pathogens. Agoutis, however, rescue a fair number of seeds by carrying them up to fifty meters away and burying them, without damage. The big-headed rodents have terrific ears and will travel more than thirty meters to retrieve a fruit that thuds onto the forest floor.

Smythe found that when no large seeds are falling in the forest, agoutis survive on those they've buried. But they don't remember where every seed is hidden, or they die before they can dine. These seeds have the best chance of maturing to seedlings. Trees like *Astrocaryum*, *Oleiocarpon*, and *Spondias*, whose seeds have an outer layer of sweet pulp that's attractive to agoutis and other frugivores, are better suited for this strategy than others. Agoutis do *Astrocaryum* one extra favor by peeling and eating the orange flesh from its palm nuts, which are riven with the destructive eggs and larvae of bruchid beetles, before their burial act.

"It seems that it may be an adaptive advantage for trees having [large] fruits," wrote Smythe, "to fruit at as nearly the same time as possible, for, if they did not, then the agoutis would no longer benefit from hoarding food and the seeds would not be dispersed." That is, if there were fewer fruits around at any one time, the agoutis wouldn't bury any for later. Smythe also proposed that if trees staggered their fruiting so that the rodents never went short, the agouti population would increase "to a level where they would destroy a larger proportion of the seeds produced."

Based on Smythe's study, Bert Leigh has suggested that the mild-mannered agouti, as common on BCI as rabbits in a New England field, may actually be a keystone species — upon which many other organisms depend — for the preservation of tree diversity in the Neotropics.

Adler's islands generally lack agoutis (they occasionally show up but never establish populations because the islands are too small), but his data suggest to him that spiny rats play a similar role in smaller forests. The rodents are abundant, they eat a lot of seeds, and they do so in "scatter-hoarder" fashion. That is, they hide seeds in leaf litter, which saves them from predation by bruchid beetles. If a rat forgets where it has hidden a seed, the seed has a chance at survival.

"Rats have a say in which trees will be successful or not," Adler told me, back when we first met. "They help determine the complexion of the forest. They like to hang around treefall gaps, areas of forest disturbance. There's more new growth in the gaps when rats are present. Rats could also serve as a very important source of mycorrhizal inoculum" — the spores that establish a new fungal colony.

Mycorrhizal fungi are a big topic in ecology these days. The fungi take carbohydrates from the plant roots they grow around, and in exchange they help the plant absorb minerals. Researchers are looking at the fungi's many forms to determine why some plants have them and others don't, and how the fungi spread through the forest.

How could rats move mycorrhizal fungi around the forest? Adler directed me to Scott Mangan, a student at Oshkosh who was finishing his master's thesis on this very subject and who happened to be sitting next to Adler, in the dining hall, as we spoke. "Scott is the world's expert on rodent mycophagy in the tropics," Adler said. We all laughed, knowing it probably didn't take much. The field was small and young.

"Rodents disperse mycorrhizal spores, which are underground," explained Scott, who was boyish and intense. "Are they smelling it? Or are they digging for roots and they come upon it? We don't know yet. We do know that they're eating only those spores associated with fruiting bodies — underground mushrooms, basically — from which the rats get nutrients.

"The more rats an island has, the more mycorrhizal fungi they eat. This seems obvious, but it's because there's more competition for fruit on those islands, and the rats turn to fungi for nutrition. The fungi's growth is triggered by soil moisture, so it's more available in the wet season. The fungi pass through the rat, and the spores are still viable when they come out. They may be washed into the soil by rain — possibly."

Scott smiled. If this particular mycorrhizal story panned out, it would be yet another example of mutualism — an interaction between two species in which each positively affects the other.

Several weeks passed, and the lake dropped another twelve centimeters. Two spiny-rat islands were now connected to mainland peninsulas. Rats that emigrated were now, presumably, competing for fruit with other mainland creatures — coatis, collared peccaries, pacas. They could fall victim to predators they'd never before encountered. Things could go any number of ways.

I went out one day with Paul to do islands 8, 5, and 53. His usual boat was on the fritz, so we took Boston Whaler 28, which had a pushbutton starter. Halfway out of Laboratory

Cove, Paul killed the engine to adjust the fuel line, and then the boat wouldn't restart.

Lacking a paddle, we drifted to shore in the light breeze. We jumped out of the boat and immediately sank past our ankles in mud, an unpleasant aspect of the ever-dropping lake. We tried another boat, number 3. Paul pulled the starter cord, and the engine nearly tore off its mount.

By now, it was nearly eight-thirty. "This sucks," Paul said. "The traps are open and the rats are in the sun. They're not gonna make it." Scouting the dock for a boat that was in working order, Paul spotted number 31, which was off limits to anyone but the scientists working on a multiyear butterfly-migration census. "Go ahead and take it if you've got animals in traps," said a bat researcher skulking around the dock.

That was all it took. Paul pulled the cord, and we were off. We dashed through island 8, which had been reconnected to Buena Vista Point for two weeks. The rats seemed fine. En route to the next island, the boat engine sputtered, then died. "Now what?" Paul said. He removed the engine cover, fiddled, yanked the cord. Nothing.

But luck hadn't totally deserted us. Within ten minutes a fisherman, out with a teenage boy, appeared. The youngster tinkered with the engine's timing. Offering occasional encouragement, the fisherman pulled a beer from a cooler, drained it, then another. The engine needed more sophisticated help, it turned out, and the fisherman towed us back to BCI. During the ten-minute ride, he finished another beer and started one more. "Wow," said Paul, after sending him on his way. "Four beers and it isn't even eleven o'clock!"

Jaime, a boat mechanic, showed Paul how to bypass the start button of boat 28 with jumper cables. Touching black to red and red to black, Paul gunned the engine and we sped toward the next island.

It was twelve-thirty now. As Paul and I raced up the shore to the first transect, his dread was palpable. "Goddammit," he said. The first rat was lethargic, obviously near death. Paul didn't even bother to weigh her: confining her in the pillowcase would be too hot and too stressful. He gently removed her from the trap, using his bare hands, then read off her toe markings. I recorded the four-digit number and solemnly printed, "Heat stressed." He laid her in the shade and said, "Let's go."

The next rat was slightly more lively but hardly in good shape. To make matters worse, she was pregnant and new to the Adler traps. Paul didn't want to stress the rat further, but a sense of obligation to Adler's seven-year project compelled him. He cut her toes. Blood spurted in a thin red line. The rat squealed pitifully.

"I *hate* to do this," Paul said. "They get so vascular when they're pregnant." We walked along transect B and soon came to another weakened rat, huddled against the cage side. "He ain't gonna make it," Paul said, gently placing him in the shade, not even bothering to read off his toes.

We blamed the boats for the dead rats. The boats didn't work because they were stressed by overuse, and they were overused because so many were broken, laid low by transmissions and motors strained in the hydrilla, by running into stumps and maneuvering around islands ringed in mud. There weren't enough boat mechanics, parts didn't come fast enough from the city. Was STRI insufficiently supportive? Did Oris, the island's scientific coordinator, know the extent of the problem? These were the questions we asked as we trudged along on death patrol.

It was after two when we reached the last island, tiny number 5. Among the four rats in traps — he never caught many here — was the dominant female. "She's large and she's often caught, so she's probably breeding a lot," Paul said. Now she was dead, slumped against the wire mesh. "There aren't going to be any rats left!" Paul shouted. Usually his voice was flat, midwestern, and imperturbable. Now he was thoroughly riled. His hair stuck straight up; he muttered to himself.

Every time we came to a dead rat, Paul sighed, overcome with sadness. He gently removed the dead animal, bare-handed. The bloody pillowcase was on the ground. I opened the logbook and he read off the toe markings: eight, four, eight, four. He droned on: Male, scrotal, adult. D.O.A.

7

Higher Primates

IT WASN'T ALL HARD WORK and drudgery on BCI, nor was
it all intellectual striving and rigor. After about a week on the
island I discovered that residents enjoyed a lively nightlife,
centered — as I had suspected — around the laboratory bar. I
got my first glimpse of this on what had been, for me, a quiet
Friday night.

Looking for information on dung beetles in the herbarium, I heard music coming from the lounge. I pulled open the
door, feeling a little sheepish. I'd been here only in the daytime, when residents flipped through old magazines or made a
quick slice of toast. Now the room was dimly lit, and Morphine droned from a boom box. One of the monkey assistants
in a tank top and cutoffs was unceremoniously mixing frozen
tamarind juice with Seco, the local rum. Over by the window,
a German in flip-flops and a bat T-shirt rifled through *Sein-
feld* videotapes, which were sent from the States by someone's
sister. Two field assistants briefly debated which episode to
watch, the one where Jerry gets discounted dry cleaning or the
one where Elaine runs out of contraceptive sponges.

They decided on the latter, though they'd both seen it before. The German stopped the music, and the assistants settled
onto the stained yellow couch. Island lore said chigger bites

came not from working in the forest but from sitting on these three sagging cushions. Still, the couch was prized real estate.

I was put off at first by the lounge's moldy smell, but I got over it quickly: this was the island's social center, the place where biologists held meetings, carved Halloween pumpkins, sewed Carnivalito costumes, staged a yearly auction, and conducted the early stages of courtship. In the corner kitchen, residents sick of dining-hall fare cooked private meals.

It was here, too, that the habits of the absent-minded scientist ran headlong into the habits of the anal-retentive nerd. Irked by the slovenly state of the sink and stove, day-visiting biologists — mostly Panamanians, who cooked their lunches here — and resident biologists — mostly Americans, who cooked at night — left ugly messages for each other on the blackboard, in both Spanish and English: "Pigs, clean up after yourselves!" It was the only bit of tension between Panamanians and gringos that I'd ever witness.

The bar itself, an oasis of serenity, was on the balcony. But there was no actual bar there, in fact. Beer came from the battered soda machine in the downstairs hallway. Residents imported their own hard liquor (or ethanol, as the dorkier scientists called it) from town as needed. On nights when a beer shipment came in, there would be a case of HB or Soberania on ice, free to anyone who'd helped unload. Otherwise, they cost fifty cents.

The bar's physical accoutrements included two benches, a few plastic chairs and one rocker, arranged around a battered low table. Geckos darted in the overhead dimness. On this, my first visit, I was taken aback to see a nearly ten-centimeter cockroach, *Blaberus giganteus*, crawl along the stuccoed wall, its antennae bobbing languidly.

These monsters are fairly common on BCI (I once found one on my pillow), and like cockroaches everywhere, they inspire loathing. But *Blaberus*, above all other cockroaches, seems actually to deserve its bad reputation. The creatures fly, they eat leather hiking boots, and they've been known to gnaw on distal body parts. In the days of sailing vessels, sailors in certain latitudes used to protect their fingernails and toenails from *Blaberus* with leather coverings.

The cockroach quickly scuttled away, unremarked upon by anyone, and I took a closer look around. Chrissy, dressed in a gauzy pink skirt, sat smoking next to Paul, whose hair was at full nocturnal attention. Sabine Stuntz, the German doctoral candidate, sprawled in a miniskirt and low heels. Field assistants filled in the gaps. The group was covering familiar conversational ground, planning a going-away party for a long-term resident. Would there be a special drink? Who would do the shopping in Panama City?

It didn't take me long to learn that BCI's residents sought any excuse to spin their dance tapes and plug in the blender. They honored every conceivable American holiday and some Panamanian ones as well. They threw parties to celebrate arrivals, departures, birthdays, and even the renovation of buildings.

Within a week or so, I felt fairly comfortable in the bar, and soon I was visiting like everyone else — when I needed a cold beer, when I wanted a little company. Depending on who was there and what time you went, the place could be either wildly stimulating or deadly dull. The younger residents stayed the latest, as a rule, and many of the older scientists, Bert included, never came at all.

I sometimes came to the bar early, to drink a gin and tonic with Bret and discuss the news of the day. One afternoon, after I'd spent the morning counting butterflies on the lake, we took our usual seats and I asked Bret if it was hard to keep in mind exactly why he was here. Was he ever discouraged by the more arcane aspects of his study? Did he ever ask himself if the drudgery was worth it?

The drudgery was necessary to get a Ph.D., Bret said, but I was looking at it upside down. "For me, getting a Ph.D. is a mechanism to do what I like — observing things in the forest.

I like being able to find tents. I like being a naturalist. Going for a Ph.D. is a justification for me to spend years thinking about these organisms and seeing parts of the world I'd never get to see."

Bret could have fashioned a thesis around any of several hundred Neotropical vertebrates, but he chose bats for one very particular reason: they were little known. And they were little known because the technology that lets researchers look into their private lives was only recently developed.

"So many questions about other organisms have already been answered," Bret said as I cut another lime and refreshed our drinks. I'd bought the limes in town, but in a few months I'd grab them for free, off a tree at the end of Miller Trail. The tropical perks were beginning to add up.

Intrigued by the idea of a wide-open field, chock full of bat and other mysteries waiting to be solved, Bret chose that path. "Can you imagine what it must have been like for Darwin and Wallace?" he continued, with marked enthusiasm. "To land in the tropics and see so many things that no one had studied before? To be confronted with so many questions!"

Studying the *Angraecum sesquipedale* orchid of Madagascar, for example, Darwin noted the flower's twenty-eight-centimeter-long corolla tube and hypothesized that "some huge moth, with a wonderfully long proboscis" must exist in order to pollinate it. Forty-one years after Darwin made his prediction, the moth was found, and suitably named: *Xanthopan morgani praedicta.*

The tropics were full of such weird adaptations and convoluted relationships between disparate species. Early naturalists had the thrill of discovering some of these organisms for the first time — water lilies the size of tractor tires; the extravagantly plumed bird of paradise, which constructed sunlit forest stages upon which to enact its elaborate courtship dance; army-ant bivouacs a million members strong.

Today's scientists can still be thrilled, but their surprises usually come from looking at these creatures' inner workings: *Heliconius* caterpillars, for example, that eat only the species of passionflower whose cyanogenic glycosides they can detoxify and convert to a source of nutrition; passionflowers that retaliate against these herbivores by creating false egg spots on their young shoots, a ploy that persuades female butterflies to deposit their ova elsewhere; fig wasps that anticipate the arrival of a competing female inside their fig and preemptively produce multiple sons, to outcompete her offspring.

Bret went on. "I'm even jealous of people like E. O. Wilson and Robert Trivers, working thirty years ago, and the things they had to think about — big issues like sociobiology, island biogeography."

Still, the mysteries of nature, for Bret, presented unbounded opportunity. And though he complained about the fragmenting of questions into ever-smaller parts and the hyperspecialization that went along with that, he wasn't truly discouraged. The forest, in all its minute and marvelous complexity, compelled him to think. That was why he was on BCI. As Stefan Schnitzer, an equally self-aware researcher, would soon tell me in justifying his years of toil on the island, "Knowledge, and its pursuit, is beautiful in and of itself."

Bret and I went up to dinner, but I kept thinking about the other residents. Why were *they* here? Bert, I surmised, was here to think and to write, in an environment that inspired him and made few demands. The older researchers were invested in complex long-term projects and had burning questions to answer — about secondary compounds in plants, about the coevolution of figs and fig wasps.

Chrissy, however, was a different matter. I began to see her, not without appreciation, as something of a careerist, someone interested in making her name in a crowded field. The big picture, where her puzzle piece fit into the larger tropical schema, held no intrigue for her; it didn't really matter.

I thought about my mornings with Paul, on rat patrol. Like so many field assistants, Paul didn't know the finer points of the project that he kept up and running. He did all the daily gruntwork, but he had none of the intellectual satisfaction of either designing the study or analyzing the data collected over so many grueling months. For the moment, this seemed fine with him. Being on BCI gave him face time with well-known scientists, a way to explore career opportunities, and an exotic line on his résumé. Other field assistants were conducting their own research projects on the side, learning to design and carry out experiments in the field. Under optimal conditions, the field assistant and the field assisted were, in evolutionary terms, obligate symbionts: each benefited the other.

When I first visited BCI, field assistants were a minority on the island. Now, almost a decade later, they outnumbered graduate students, postdocs, and tenured professors combined. Among field assistants, comely young women — hired by male professors — outnumbered men by about four to one.

According to those who engaged them, field assistants were now an absolute necessity. "Only ten to fifteen percent of ecology grants get funded by the NSF," Lissy Coley told me. She and her husband, Tom Kursar, visited BCI every year. Collaborating with Merck Pharmaceuticals, they employed several assistants in their search for cancer-fighting compounds in plants.

"Scientists who do cellular and molecular work get more grant money and more respect," Lissy continued. "It's harder, today, to get funding for small, short-term ecology projects than for larger projects. So you devise a bigger question and attack it better. That means hiring assistants."

The trend was a reflection, as well, of the academic hiring crunch back home. Understaffed and underfunded American universities didn't want their professors absent eight months of the year. Principal investigators who wanted to keep their research projects on track, so that they could continue to publish the papers that kept them employable, needed bodies in the field, collecting data year-round. Undergrads like Paul were happy to oblige.

For those who stayed too late at the bar, mornings on BCI could be brutal. At dawn, the island was noisier than midtown Manhattan. Howler monkeys bellowed, birds screeched, mystery animals dropped fruits onto the tin roof of the dorm, coatis snuffled through the dry leaf litter just outside the windows. If that racket didn't wake you, then the 6:00 whistle of the *Jacana* as it left for Gamboa would.

When the launch returned at 8:30 A.M., it discharged a motley collection of scientists, laborers, and administrators. Like some other social animals, these higher primates were organized in castes. After trudging across the dock as one, they immediately went their separate ways: the Panamanian white-collar workers to the office, blue-collar workers to the boat dock and maintenance sheds, postdocs to the lab, *guardabosques* to the forest. Guardabosques are, literally, forest guards: rough-and-ready Panamanians who wear camouflage, carry sidearms, and valiantly patrol the island for poachers. (Looking for snares and listening for gunshots, they sometimes catch two or three poachers in a night, or none for an entire month.)

Twice a day, all castes reconvened in the dining hall: Panamanian workers took the table by the lake window; guardabosques sat near the toaster oven; scientists, the largest group, filled the long table near the buffet. Among this last group were subcastes: STRI scientists, professors, postdocs, doctoral candidates, master's students, and field assistants. The young Germans made up their own clique, with a look and a political consciousness straight out of the 1960s. The men had long hair. Everyone wore tie-dye and smoked. The Germans had the best photography equipment, they drank the most beer, and they were always eager to swim at night.

Nonscientist tourists, who clomped onto the island two days a week, ranked the lowest in BCI's pecking order. Residents never mixed with them. Years ago, the beer jefe hung up a sign that said, "Beer: 35 cents. Insert dime first." Tourists fed the machine the dime, then the quarter, and they got their beer. None of these outsiders left the island knowing that the price of beer was a quarter. The extra dime was just gravy.

During my first visit to BCI, I got the impression that it was a haven for weirdos — for socially maladapted scientists. I was enthralled by these residents. They were smart and direct, and they didn't suffer small talk and pleasantries gladly.

The island indulged such social quirks. Here, nothing raised an eyebrow. At dinner you could practice frog, bird, and monkey calls. You could spend an entire day sitting under a tree. You could hide in your lab for a week, speaking to no one, and not even be missed.

Collecting data alone in the woods, then analyzing them and writing it all up, was an arduous and solitary pursuit. One was either driven away by the solitude — as Chrissy might be — or one adapted. In *Naturalist*, E. O. Wilson wrote that to be good at something, the biologist must have the capacity to absorb himself in a group of organisms or an environment, to forget about daily life and be away from people. He must, in other words, be an introvert.

The oddest and most interesting people I met on BCI were the scientists in their forties and fifties. Were they a self-selected group? Had the gregarious drifted into other work? Perhaps the younger scientists simply hadn't had time to develop the physical and psychological tics that marked so many older scientists, who seemed painfully shy and inner-directed. Some of them could hardly bear to make eye contact. They were islands, in and among themselves.

Some of these scientists, in their own dowdy way, I found glamorous. They had dozens of papers under their belts, their work was cited, and they had a deep knowledge of the forest from spending so many years under its sheltering canopy and around BCI's dining table, a station of the tropical biologist's cross. Unlike the residents just starting out, with their narrowly focused studies, the senior scientists had a grasp of the big picture. Their projects were long-running and long-ranging. The STRI scientist Joe Wright, for example, ran the Environmental Sciences Program, the Psychotria project, the canopy crane, a forest fertilization experiment, and studies of the canal's watershed, mammal abundance, and palm regeneration. An army of field assistants, whom Bert referred to as Joe's slaves, made it all possible.

Wright's projects were ecological: he looked at the tropical forest as a system and tried to figure out how it functioned and maintained itself. Allan Herre, another STRI scientist, was more interested in general questions about evolution. To study the phenomenon of cospeciation, he examined the relationship between fig trees and their exclusive pollinator, fig wasps. The two organisms were mutually obligate: when fig trees radiated into numerous other fig-tree species, so did their mutualist wasps. But how did the mutualism evolve? And why does the fig wasp lay eggs on some ova of the fruit but leave others to develop into seeds? What kept this mutualism from verging into parasitism?

For a beginner like me, it was easier to see the value in Wright's and Herre's brand of basic research than in the isolated projects of the young professionals. By looking at what was out there and what maintained it, Wright's studies had obvious relevance to conservation. By examining the engines of evolution, Herre's studies were telling us, ultimately, about the forces that created us.

In between these projects lay a lot of open territory, populated by BCI's doctoral candidates and postdocs. Their work would fill the pages of the professional journals, from *Nature* and *Science* on down, and their findings would stimulate new questions.

Adding to the continuum of knowledge was an admirable thing. But every now and then, something truly fortuitous would come of a younger scientist's work. One of these tiny puzzle pieces would fit, miraculously, into another, and a part of the puzzle that no one expected to be connected was now that much closer to completion. The tropical rain forest, mysterious and dim, was suddenly just a tiny bit brighter.

It was the bar that gave me a sense of BCI's pecking order. A new arrival would tentatively creep out onto the balcony, someone would find out that he was here for just three months, and the welcoming arm of social inclusion would instantly retract.

Of course, different people had different criteria for deciding with whom they would engage. A person staying for six months, I was told, might befriend someone visiting for three, but a person staying for years wouldn't befriend a summer assistant.

The station had always been an in-group sort of place, where long-term residents avoided involvement with short-termers. The phenomenon is common to any isolated spot that sees high turnover — mountaineering base camps, the journalists' watering hole in a foreign city. BCI residents didn't invest energy and time in people who were just about to leave. "It's a bit wrenching," one scientist told me. "You're always left behind." Eventually, long-termers learned to avoid the heartache.

The scientists who worked on BCI for years, or who had been visiting yearly for decades, occupied the highest rung on the social ladder. Longevity brought its privileges: a better lab, a private room, one's own mailbox. Long-term residents knew how to get boats and cars, food from kitchen stores, a bedsheet free of someone else's stains. Long-termers cultivated relationships with island administrators and so had access to information — the ultimate currency of the island — that the three-month visitor never got.

In its earliest days BCI resembled an exclusive intellectual and social club. Visitors reveled in the field station's collegial atmosphere and in its simplicity, which allowed well-to-do scientists to play at being "uncivilized" or "savages," as one such visitor wrote.

Undergraduates were not allowed, only mature investigators "of recognized ability and standing." David Fairchild wrote to Thomas Barbour, in 1924, "We must prevent this institution of ours from drifting on to the shoals of popularity and the crowded state." Fairchild believed that "undesirables" could "ruin the peace of a camp."

During BCI's first two decades, women were easily excluded: they had no place to sleep. Fairchild saw to that. To Barbour he had written, "I do not believe the same curiously stimulating atmosphere can be maintained in a body of men and women that can be in a body of men alone."

"I understand that men and women are pretty promiscuously intermingled at a certain zoological station in British Guiana," wrote another concerned scientist, in 1924, "but so far the breath of scandal has been wafted away from Barro Colorado Island, and being near neighbors to a [L]atin [A]merican community where such advanced license is still discountenanced, I think we may well take good care not to get ourselves talked about."

Eventually, though, BCI gave people reason to talk. As its popularity grew, its doors opened wider. The research station gradually metamorphosed into something more akin to an army barracks, which made it socially difficult on the women who first came to work here, in the 1970s. Most of the residents then lived atop the old dining hall. Walls between beds didn't reach the ceiling, five people shared a bathroom, and everyone knew everyone's business. The island had no telephone. Reading in bed was difficult because ceiling fans hung below the light bulbs, creating a strobe effect. The sounds of sex and snoring were broadcast from room to room, and the clamor of breakfast preparations, downstairs, woke everyone before six A.M.

"The social dynamics were extreme," recalled Lissy Coley. According to other women who lived through that era, many of their male cohorts — who weren't accustomed to working with, or living among, females — lacked social grace. One biologist would seat himself next to the most recent female arrival, place a rough paw on her thigh, and bluntly ask, "How 'bout it?"

When I asked Bert about such behavior, he said, "The men weren't sexist. They didn't discount the women's ability. But some of the women were not as sexually responsive as the males wanted them to be."

As an outpost of undersocialized men with expectations raised by the sexual liberation of the 1960s, BCI was not unique. In the late '70s, when men dominated Antarctica's McMurdo Station, one microbiologist devised her own taxonomy for the Antarctic male. There were predators, who lurked at every turn; scavengers, who were usually considerate enough to wait for a signal before pouncing; and herbivores, who wanted nothing to do with women.

Couples on BCI have always come together quickly — probably because people live in close quarters, share common concerns, and are interested in one another's work. One summer, guessing the nature of male-female relationships became something of a parlor game. A couple of postdocs, in typically analytical fashion, drew up a flow chart delineating who was sleeping with whom. A solid line represented a confirmed alliance, a broken line a suspected one. A few alpha males and alpha females had the most lines, but "everyone," according to Robert Dudley, an entomologist and flight physiologist who'd been visiting BCI since the late 1980s, was doing it. "Things were pretty tense; it was a high-level feedback loop."

In the early days especially, privacy was at a premium. At one time or another, there was mating on the raft, atop the canopy tower, in the walk-in refrigerator of the old dining hall, in the labs, at the base of the Big Tree. When the new dorms went up, in 1990, alfresco sex went down. But not out. During slow times, one had only to check the room assignment board to find a vacant trysting spot.

Robert tried to persuade me that people on BCI acted in ways they wouldn't conceive of in the "real world": they cheated on spouses, they propositioned randomly; one man conducted affairs with two women at the same time without either of them knowing about the other. And the women were roommates!

For the shift in social mores Robert blamed isolation — from friends, lovers, and other comforts, and from the distractions of home. Then there was stress, and cabin fever. And keep in mind, he added, "that scientists aren't the most well-adjusted people in the world. They spend so much time developing one side of themselves that they often ignore their own social development."

"You mean they can be immature?"

"Precisely."

At a party during my stay, a married Venezuelan hit on one woman after another until he got to a blond student named Eileen. "I don't like you and you're making a lot of females here very uncomfortable," Eileen said. After that, he just skulked around the fringes of the gathering, plucking mantids off the window screens.

A professor who visited regularly got drunk in the bar one night and went around the circle, asking each woman if she thought she was a "good lay." On another night, in similar condition, he asked a married man if his wife was "available."

In fathoming the ribald behavior of some residents, Robert Dudley had perhaps overlooked one of the most important factors: they thought they could get away with it. No one back home would find out, or so they thought, and illicit relationships would end with the field season.

Are the tropics sexy? A lot of people think so. There is an idea that the lower latitudes encourage sensuality because they are

exotic and remote and because people spend more time outdoors in such places, being physically active and wearing less
clothing. I didn't buy that, mostly because it wasn't the way *I*
felt. Also, more people were not having sex on BCI than those
who were. At least that was my impression.

As a diurnally active writer, I could never be certain who was
hooked up with whom. Researchers with nocturnal study subjects usually had the best inside dope — they walked around
a lot at night. But sometimes couples were obvious, and the
entire community witnessed their happiness. Breaking up, in
these cases, was hard on everyone; it was all so public. One
night at the bar, a woman serenaded her new lover on his
birthday; inside the lounge, his old flame sat alone in the dark,
blasting Ani DiFranco on the stereo: "Fuck you and your un-
touchable face."

The stress of work and isolation got to people. At least once
a fortnight, the cry "Naked raft party!" would rise from the
late-night bar and residents would take to their Boston
Whalers. The hooting and hollering of swimmers could be
heard in the dorms until the small hours of morning. Getting
naked, though, wasn't peculiar to scientists in the tropics: at
McMurdo Station, scientists converted ice-drilling equipment
into a hot tub and residents strove to join the 200 Club, whose
members went from 100 degrees below zero, on the ice, to 100
above, in the tub.

In the beginning, it was sometimes hard for me to get scientific
information from scientists. I was an unquantified entity, I occupied a fuzzy place in the pecking order, and many residents
were paranoid about discussing their research with people they
didn't know. In an environment where ecologists compete
closely for limited grant money, they didn't want their ideas
"parasitized," as one resident explained to me. Moreover, their
field, compared to something like physics, is young. In physics,
general principles are known. In ecology, there is still a great

deal of competition to "get there" — to arrive at the great theories — first.

After two months on the island, however, I sensed that relations had more or less normalized. Residents accepted that I was here for the long haul — coming and going over the better part of a year — and that I posed no professional threat. I decided the annual auction was a good time to make my formal social debut. A much-anticipated affair, the auction was invented to rid the lab of equipment and clothing left behind by residents, to clear any beer fund debts, and to raise money for the BCI scholarship fund, which went to a deserving Panamanian grad student.

We gathered in the lounge at eight o'clock, and the items came fast and furious: eighty plastic cups with holes drilled in them, tags for plants, ropes for climbing, sunscreen, glass vials, PVC tubing, microscope slides, a box of toiletries. Everyone was drunk on tamarind juice and Seco. Every few minutes the auctioneer, dressed in a donated spangled maillot, held aloft an item — a bucket, a plastic box, a T-shirt — that had previously belonged to someone named Steve. The women went wild; the bids erupted. Who the heck was Steve? The level of sexual innuendo rose steeply with every mention of his name. Steve wore the T-shirt? The bids went up. Steve touched the box? Steve pissed in the bucket?

Finally I asked a resident, sotto voce, "What's with the cult of Steve?" "He was a heavy-drinking, Tarzan-image, tree-climbing womanizer" came the reply. I bid on one Steve item, a straw hat, but was happy to drop out when the price got steep.

At around midnight I made my move, offering my services as field assistant for a day. Those with whom I'd worked offered testimonials. The offers rose quickly from 50¢ to $7.50. I was bought by a quiet plant physiologist named Dan, who wanted me to record data as he measured a couple of thousand seedlings in an experimental plot on the far side of the island.

I woke the next morning with a wicked hangover. On my

desk was a large green T-shirt with holes in it, a pair of very short purple-flowered shorts, a rubber ball, and a large bag of black crepe paper. Apparently I'd shouted out the night before, "I have plans for this!" — but I had no idea, now, what they were.

The auction marked a turning point for me. By then I was perfectly comfortable in the forest, on the trail and off. Working with different researchers, I was developing different search images: for bat tents, tungara frogs, spider monkeys, butterfly eggs, leaf-cutter nests, and more than a dozen different plants — their fruits, seeds, and leaves.

Having numerous guides gave me numerous perspectives on the forest and on the experience of being a scientist. I was beginning to think that as field assistant without portfolio, I had the best job on the island. I skipped between projects, never bored. Out of necessity, my education was general, in a world of narrow specialists. But I took what I could get.

Like the other field assistants, I rose before dawn, tucked my pants into my socks, and filled my water bottle. In my pack I carried a map, a compass, a knife, and a sheet of plastic to sit on. If I was going out with Paul, I brought extra water. If I went out with Chrissy, binoculars. Rafael, butterfly net. Bret, headlamp.

In the months to come I'd do some odd things. I spent a few afternoons using a turkey baster to empty and measure the water that filled tree holes. These holes, in which damselflies, *Megaloprepus coerulatus*, lay their eggs, are the deeply fissured bark of trees lying horizontal on the ground. They can be a few centimeters deep, or half a meter.

One afternoon I helped measure several thousand tiny *Brossimum* seedlings and feared that I'd never stand erect again. Another day, I wrecked my hands tearing open the tough seedpods of *Platypodium elegans*, a graceful tree with a braided trunk, so a doctoral candidate could analyze their DNA. I signed on for an all-night mammal census, in which I'd sit in a tree platform with a night scope, but somehow the plan never came off. I learned to sex butterflies, chasing them in a boat across the flatland of Gatun Lake and then the wave-tossed Caribbean Sea. I clipped the toes of tungara frogs and spiny rats. I climbed saplings to reach leaves that another researcher had knocked out of the canopy, with a shotgun, so she could track paternity among the species. I captured euglossine bees in nets sized for goldfish, luring them with scent strips tacked to trees. On another project I left these splendid bees alone and simply watched them, for hours on end, as they pollinated an equally splendid flower.

Despite these many hours in the field, my understanding of the forest was relatively superficial, a point that was driven home to me every time I attended a Bambi. On Wednesday nights I'd sit in frustrated silence while researchers, using language that I deemed intentionally obscure, presented their data on water transport or the species composition of forest plots. I had a million questions about these projects, all too dumb to ask. I didn't understand the statistics, I didn't understand the implications of the investigations. During my worst moments I damned tropical biology as a black-art discipline and scientists as high priests of esoterica.

How, I wondered, did investigators sustain their interest in studies so narrow and obscure, studies whose payoff, if there was even to be one, was so far removed? Certainly there were easier careers to follow. And more sophisticated, better-paying kinds of science, too.

Instead of learning about evolution in the tropical forest, BCI's mud-spattered residents could be learning about evolution in a test tube. Every week, it seemed, neuroscience and molecular biology made a breakthrough — a new receptor for painkillers, a possible treatment for cancer. But the world, alas, did not eagerly await word of the minimum required distance between tungara frog breeding ponds.

And yet a working-class pride infused the resident biologists. They cultivated a spunky disdain for lab jocks, for overfunded molecular biologists, and for commercially viable research. They, the field biologists, were working in the tropical rain forest! They were examining living organisms in their natural context! "Fruit-fly geneticists wouldn't even know where to find their particular *Drosophila* species in the wild," one entomologist told me with an exultant sneer.

Bambis typically began with the project's general outline. Presenters rarely mentioned what compelled them to study the organisms they studied, or why their research mattered. I had to seek such information on my own time.

Scientists comfortable with eschewing utility proclaimed the value of knowledge for knowledge's sake. Others, oblivious to public opinion, announced they did basic research because it was "fun." They liked science, they liked the tropics, they liked climbing trees and capturing animals and putting their minds to thorny questions. Robert Dudley, my private anthropologist, told me, "I work on BCI because the diversity here is inspirational, and because the systems I've worked on for thirteen years — migration, orchid-bee flight performance, and frog vocal sacs — are still wide open."

Eventually, though, just about every scientist on BCI, even Robert, got around to saying that his or her work was, at some level, connected to conservation — that to save species we had to study them. It sounded like a party line, though, and I didn't believe it. It seemed to me that a lot of residents didn't either, not deep down. In the context of vanishing rain forests, how could an understanding of how ants parasitize caterpillars matter? Or knowing the forest's fruit-to-frugivore ratio, or the amount of green material that arthropods consume? At the end of the day, what did such knowledge add up to?

Clearly, between doing the work of biodiversity — exploring, cataloguing, and explaining it — and acting on its behalf lay a disconnect the size of the canal. Bret was one of the few biologists on BCI who acknowledged this schism.

"People here are fooling themselves if they think their work is related to conservation," he told me one night as we searched the forest for a fallen transmitter. "Most of it isn't. If they really wanted to conserve habitat, they'd work to redistribute the wealth and get slash-and-burn farmers, ranchers, and loggers to stay out of the forest. Conservation is a political problem. The powers that be won't be dissuaded from destroying the forest by an explanation of its mechanisms." Conservation wasn't about studying mutualisms, Bret continued, but about forcing Shell Oil to raise gas prices until they reflected their true cost to the environment.

To Bret, measuring how long a bat flies in search of food wasn't an esoteric pursuit but a means to an end — not to an understanding of phyllostomid bats, but to training his mind and earning his degree. "With a Ph.D., I can talk about ecology and conservation and people will be more likely to pay attention to me," he said, pawing through some leaf litter. "And conservation is the only goal for me. I don't believe those people around here who say they're helping to save the rain forest by studying some minute interaction between two organisms. But there is value in that kind of knowledge because it's a way to start talking about what's at stake."

Bret was quick to acknowledge that what he learned about tent-making bats would not help to conserve them. "But I like to talk about bats," he said, "and forests would look very different without them."

Forests would look different because bats are frugivorous: they disperse a lot of seeds, particularly those of pioneer plants, which recolonize areas that have been recently disturbed or clearcut. Bats also pollinate or disperse the seeds of bananas, breadfruit, plantains, mangoes, figs, guavas, avocados, cashews, cloves, chicle, balsa wood, and kapok. Without bats, there would be no agave. No agave, no tequila.

Highly successful mammals, bats live all over the world, except at the very highest latitudes. They roost under leaves and rocks, inside logs, termite nests, hollow trees, caves, and attics. Bret continued, "Because they live in such a wide range of niches, consume a lot of different foods, and react to habitat disturbance, bats also make good indicators of ecosystem health."

"So bats can be an indicator species?" I said, averting my headlamp from his eyes.

"Yes," Bret said, "if you have good baseline data on them before human disturbance occurs."

I smiled to myself in the dark. Almost every scientist I met told me his study organism was an indicator species.

One Wednesday, the *National Geographic* photographer Franz Lanting arrived and usurped the regular Bambi slot. Lanting was a shaman to a particular tribe here — the tribe of aspiring nature photographers and potential *National Geographic* feature subjects. Some wanted to show him their own slides, others wanted publicity for their studies. A mention in the *Geographic*, while not a peer-reviewed journal, was still pretty close to gold for any scientist who'd grown up reading it.

Lanting was here to shoot a story on biodiversity. He spent the afternoon photographing a three-toed sloth, which hung its arms limply around an assistant's neck and stared, with its characteristically blank expression, directly into the camera's lens. Sloths move on lianas between trees about every day and a half. Members of the order Edentata, they punctuate their week with a single trip to the forest floor to urinate and defecate. This operation takes about thirty minutes and includes the digging of a small hole — with their stubby tail — for the burial of excrement.

In 1975, two researchers estimated BCI's sloth population at between nine and ten animals per hectare (a figure some consider high). But the sloths' white fur, tinged green with algae, makes them cryptic, and they aren't always easy to find.

Lanting located his sloth the easy way: by paying Bonifacio de Leon, the island's biological technician, to produce one.

After dinner, residents assembled in the conference room to see Lanting's slide show: endangered lemurs in Madagascar, rare parrots in Central America, emperor penguins in McMurdo Sound. The slides were spectacular, and they had the effect that Lanting intended: they persuaded us that he was for real, and that we ought to cooperate with him. "If you know of something interesting, something that's repeatable for the camera, let me know," Lanting entreated the crowd.

Bret, sitting next to me in the front row, found Lanting smug and arrogant, but he believed in the photographer's project. "A lot of people are going to read this story," he said. "It's important to raise awareness, to let people know what we're losing."

Some of Lanting's best pictures had been taken in Madagascar, in a northern parkland called Ankarana. Almost in passing, Lanting said that since the time of his visit, industrial emeralds had been discovered in the forest and ten thousand miners had moved in.

"I can't believe they're mining in Ankarana," Bret said after the lights came up. He was slumped in his chair, shaking his head. "I was there two years ago, and I'm sure that Heather and the guide and I were the only ones in the entire beautiful park. This is so depressing."

Like many a park in developing nations, Ankarana was protected on paper only. When a value with more immediate payoff than biodiversity came along, the "protections" evaporated and lemurs lost out.

Now Bret was frowning. His mood, which had been ebullient at the beginning of Lanting's show, would remain dark for hours. I was struck by how deeply he felt the loss of the park. I hadn't expected an emotional reaction from a sophisticated scholar of biodiversity. BCI's residents were intellectuals, after all, devoted neither to activism nor to consciousness-raising but to understanding the machinery behind the jumbled green façade. And yet the loss of a forest, a distant relative of his own forest, affected Bret profoundly.

Here, then, was another peek at the motivations of today's biologists. People like Bret study the tropical forest because it is a spectacular thing, and because that spectacle is, on our watch, becoming rare.

8

The Rainy Season

IT WAS THE RAINY SEASON NOW, and all the island streams were flowing, with pools big enough to sit in. The island looked more than ever like a rain forest — shiny green and lush, with sweet-smelling flowers and the sound of trickling water. I saw fruits I'd never seen before — *Coccoloba densifrons*, with its small yellow berries, and *Tontelea richardii*, smooth, brown, and coconut-sized. The petals of newly flowering trees carpeted the trails.

It was water and warmth that made the tropics tropical, and the onset of the rainy season that cued so many of BCI's plants to flower and fruit, setting off rhythms that reverberated through all trophic levels. Of course the onset of the dry season sent signals as well, but the rain, well, the rain was *it*: the reason all of us — creatures who studied and creatures who got studied — were here. It was the element that literally and metaphorically set the whole thing off.

Now, rainclouds shuttered out the sun for most of the day. At times, so little light reached the forest floor that it was difficult to read the words on a map. The island was gloomy sometimes. Depressing. That was when I started to notice the myriad odors of rotting wood and mold. Strange fungi sprouted on the trunks of trees and on spongy logs. There were whitish growths that looked like bean sprouts, blobs of orange with the consistency of jelly.

Tree falls would soon be at their peak, when soils and trees were sodden and strong winds blew. Slowly, all that dry leaf litter cluttering the forest floor would disappear, churned once again into available nutrients.

It was the season of mangoes. Large trees, non-native to Panama, dropped hundreds of the ripe fruits near the boat dock in Gamboa. Everyone on the island knew how to twist out the fruit's seed, score the inside flesh into a checkerboard, then turn the hemisphere inside out to bite off the juicy chunks. The only problem was opening our pocketknives: the constant moisture welded them shut.

Convective thunderstorms punctuated our afternoons and evenings, and most researchers planned their fieldwork for the morning. From the lab balcony, we watched masses of gray and black clouds move in from the northeast. First the mountains disappeared, then the lake, then the dock. Scientists walked around dripping wet, grinning. Sometimes the rain was soft; sometimes it came down in torrential sheets that turned gutters into geysers, culverts into cataracts. At those times we'd take plastic trays from the dining room and ride them, hydroplaning, down the steep path between the labs.

Indoors, all this moisture took another kind of toll. Bedsheets felt perpetually damp. My pillow began to smell rank, reminding me of the hundred sweaty heads that had lain here before mine. Chrissy told me to store the pillow in the dry closet by day. This helped some, but sending the pillow through the washer and dryer soon became routine.

Muddy, wet clothing festooned every dorm's balcony. Ceiling fans ran day and night to keep the moisture level down. No one discussed the environmental implications of this profligacy. The clank and rattle of backpacks bouncing around the dryers joined the nocturnal chorus of frogs around the laundry room.

Lying in bed during a massive storm was heavenly — that is, if I didn't have plans to go out with Chrissy at five-thirty in the morning. The rain beat upon the tin roof like a snare drum. Lightning flashed through the drawn shades. I thought about bat researchers cowering under trees with their nightscopes, and I snuggled deeper under the sheet. Once or twice I was actually cold. I shut off the ceiling fan and curled up under a towel. The rooms had no blankets.

I left my boots outside my room now, as their soles were inevitably caked with *barro colorado* — red mud. It was oxides of iron and aluminum, which weren't taken up by plants, that gave the island's soil its red color. I was as strict as a Japanese about excluding shoes from my room, and I bought a woven bamboo mat, for ninety-nine cents on Panama City's Avenida Central, for psychological reinforcement.

Some residents who spent their day in damp clothing developed fungal infections, which gave them a strange odor. After a few hours in the field, Chrissy smelled, not unpleasantly, like bran cereal. She treated the condition by washing her body in Head & Shoulders. I was spared this fungus, but others invaded my room. Gray-green mold coated my wallet. The dollar bills were loathsome to touch. My rubber boots were moldy, and my leather sandals too. Rubber Tevas took on an especially noxious odor this time of year. Tampax not stored in Ziplocs exploded, on their own, like milkweed pods in autumn.

One rainy night, a brown-bearded, brown-haired scientist walked through the dining hall with a reddish beetle on his sleeve. "It's got a larva in its mouth," he pointed out. The larva's last segment — called the pygidium — was just poking out. "Do you know what it is?" I asked, referring to the beetle. "No," he said with finality. He and the beetle dropped off their dirty dinner tray and walked out to the lab.

That was my introduction to Robert Stallard. He was insect-free at breakfast the next morning, and I asked him what he was doing here. "I work for the United States Geological

Survey," he announced. "And I've been coming to BCI regularly since 1984." When he started visiting the island, he told me, there were "some very arrogant, unpleasant people here." He dealt with them curtly: "I tried to figure out if they knew what they were doing. I'd ask them flat out."

How did they take that?

"It didn't really matter," he said. "Because after all, I *did* know what I was doing. *I'm* the one who had a job."

Stallard, at forty-six, had some odd tics, which thrilled me. He was the classic BCI weirdo — smart and nerdy, hyperfocused on work and socially awkward. Stallard rarely made eye contact. Conversations with him tended to end abruptly. When seated, he compulsively bobbed his foot. Having a drink with Stallard and Bert, who had many of the same habits, was a festival of scratching, toe tapping, and other expressions of nervous energy.

It had been raining heavily lately, and huge volumes of water poured down through the lab clearing. The storms seemed wildly destructive: hour after hour, leafy branches went swirling down the culverts; soil poured down the hillside into the lake, which was stained dark brown for several meters offshore. Would the island eventually just wash away?

Stallard was the perfect person to ask. As a biogeochemist, he studied the movement of chemicals from the bios, or living parts of an ecosystem, to the geos, or nonliving parts. He knew all about runoff and sediment and what these processes said about events upstream. If rivers left chemical signatures — and they did — then Stallard was one of the most qualified graphologists around.

"The chemistry of rivers can be related to the geology of their catchments," he said when I caught up to him in his lab. It's a neat system, made possible by gravity: mountains and hills erode, then their sediments move downhill and eventually get deposited in streams, where, if Stallard's been around, they get captured and analyzed. The chemical signatures of water and sediment tell Stallard about such environmental processes as weathering and soil accumulation and erosion. Basically what Stallard does is travel around the world setting up systems to sample streams — their water, temperature, pH, sediment load, and electrical conductivity, which indicates salinity.

"If you understand the underlying relationships between climate, topography, geology, soils, and vegetation," he said, "you can develop a comprehensive description of weathering and erosion." And what good is that? I asked. "Well, you could learn about the stability of land masses; you could predict earthquakes," he said. "Earthquakes kill tens of thousands of people a year, the world over."

How much of BCI disappears during the rainy season? I asked.

Stallard tapped away on his computer keyboard. "I can tell you exactly. In Lutz Ravine you've got a parallel retreat — where the slope contracts uniformly, from top to bottom of a hill — of 0.24 millimeters per year. On the plot, you've got erosion as well, but it's much, much slower: just 0.008. The difference is a factor of 30."

So the island was getting smaller, but not by much. Soil was also accumulating as vegetation decayed and as insects, like termites and ants, recycled organic matter into new layers of forest floor.

Figuring out how much water drained into and flowed out of Lutz Catchment was not easy. Years ago, Stallard set up a series of bubble gauges and meter sticks, above and below ground. He plugged his numbers into what Bert called "a wretched little formula" and was soon getting a pretty good picture of the catchment's dynamics. But an anomaly puzzled him.

Stallard discovered that the water level in the weir that impounded the catchment jumped every morning at six and then soon went back down. Troubled, he checked his gauges and discussed evapotranspiration with ecophysiologists and drainage rates with hydrologists. Still he came up with nothing.

Then one day Stallard fell into conversation with an animal

behaviorist studying BCI's birds. (Stallard made a point of talking to everyone on the island. "You never know when someone will have something interesting to tell you," he said to me.) The scientist told Stallard that each morning at six he saw Alice the tapir, a friendly rhinolike beast who daily partook of kitchen scraps, wade chest-high into the weir and empty her bowels. Then she left.

Eureka, Stallard said. He made some quick calculations based on Alice's geometry and before long had his mathematical models in order. He could now account for every millimeter of rain that entered the catchment. The currency of old-fashioned observation in sussing the mysteries of nature went up another tick.

Stallard's office on BCI was filled with computer equipment, parti-colored maps of the Canal Zone watershed, and several dozen Nalgene bottles, which he used to "sample events." Events were storms and samples were stream water. "There's a computer out there on the stream that, during a storm, sends a message to the gauge to start recording data," he explained. The gauge also signaled a technician, who responded by carting twenty-four plastic bottles to the stream and filling them with water. The full pack weighed about thirty kilos, roughly sixty-five pounds. Stallard asked if I wanted to see a gauge, and we headed out the next morning, without any plastic bottles. The sun was shining brightly.

From Zetek Trail we picked up Conrad, which James Zetek had named for his high school biology professor, who, according to Zetek, had picked him off the streets and gotten him interested in science. The trail led to a faint path between graceful *Oenocarpus* palms and down to a shimmering, shallow stream.

Behind a small concrete weir, water pooled. It was shady and cool here, with greenish light reflecting off the undersides of giant leaves. A blue morpho flitted through. With wings closed, these butterflies were practically invisible on the ground or on a tree trunk. In flight, though, the brilliant deep blue of their upper wings was so dazzling it was thought to help in evading capture. The Neotropics contained about eighty species of morpho, some of them blue, some silver, some brown. The largest morpho, *Morpho peleides*, had a wingspan of some fifteen centimeters.

"Nice spot," I said to Stallard, thoroughly enchanted. "Aren't you tempted to swim here?"

"Not really. You could get leptospirosis." Leptospirosis was a potentially fatal waterborne bacterial disease. Stallard said that a colleague of his, pregnant at the time, got it in nearby Soberania National Park.

I asked what happened to the woman, and to the baby. "I don't know," he said. "I guess I never followed up on that."

We stood next to a stack of metal shelves along the stream's edge. Inside a plastic case was the stream gauge, called a bubble flow meter, with a readout offering the date, the time, and the news that the water level was currently eleven centimeters above the bottom of the V in the middle of the weir. There was a rain gauge on the top shelf and a solar panel atop a small platform about four meters high. A copper tube led from the bubble meter to the pond.

"The meter sends out one bubble per second to the pool," Stallard explained. "The pressure of the air at the gauge is equal to the pressure of water through which the bubble rises. The gauge measures the air pressure inside the line. As the water level rises, the bubble needs more pressure to be expelled. An equation describes how fast the water flows through the V based on the water's depth."

It sounded overly complicated to me. "Why don't you just use something like a floating gauge to measure the water level?" I asked.

"It's too noisy," Stallard said. "Debris and storms screw it up." The bubble system was steady and accurate. It told researchers that, in the absence of rainfall, it was evapotranspiration — the water transpired from vegetation plus the amount evaporated from soil and water bodies — that caused stream flow to fluctuate over the course of the day. "On sunny, rain-free days, at six A.M.," he said, "the flow starts to decrease, until about six P.M., when evapotranspiration turns off and the ground water begins to replenish the flow." The loss of water by evaporation from the pond and transpiration through surrounding vegetation was far greater than the amount that simply flowed downhill.

Conrad Catchment drained most of the 50-hectare plot, a relatively flat area; it made a nice comparison with Stallard's other monitoring station, on Lutz Ravine, which was relatively steep but drained just 9.73 hectares.

In addition to monitoring stream flow on BCI and helping to make a precise topographic map of the island (something the bat researchers, who were radiotracking their animals, were rabid about getting their hands on), Stallard was also modeling the canal watershed for the Panama Canal Commission, which was keen to understand the processes that delivered or withheld water from the canal.

"Understanding and predicting global environmental change is a major concern of the late twentieth century," Stallard said as we walked back to the lab. "Atmospheric scientists have made a lot of progress in modeling climate. Now we need to know the processes associated with exchanges of water, energy, and carbon between the earth's surface and its atmosphere. We want to be able to predict terrestrial water, energy, and biogeochemical budgets over time and space."

Following Stallard along Armour Trail — he walked only slightly faster than Bert did on our forest tours — I wondered what losing the forest would mean to the canal. If the vegetation and the soil took so much rainwater, wouldn't deforestation actually help fill Gatun Lake?

The canal would be under complete Panamanian control as of noon on December 31, 1999. What the striving nation would do with Canal Zone forests, long protected by Americans, was much discussed in environmental circles. There were plenty of people desperate for land — to live on, to use as grassland for stock, as a source of timber and food. In 1952, the canal watershed, which includes land beyond the Zone, was more than eighty percent forested; today, only forty-seven percent is forested. And if El Niño events became more frequent or intense — as many believed they would, especially if human activity contributed to such extreme weather patterns — the isthmus would surely see more drought and more flooding, which would in turn increase population pressures in areas previously protected.

We were just about home now and Stallard answered my first question, about deforestation's effect on runoff, with "We don't know." I had a follow-up: Does the microclimate of a forest make rain? "No one really knows," he said, "which is why it's so important to study the forest: it's here now."

He pulled back a vine that blocked his path. It snapped back in my face. The same thing had happened this morning. I found Stallard's obliviousness to his surroundings amusing. It didn't surprise me that he didn't know what became of the woman with leptospirosis.

The dining room was crowded with visiting tourists when we arrived, and Stallard and I couldn't sit together. We'd speak again during the next week, but each conversation was nearly as constrained as our first. One morning I looked for him at breakfast and then in the lab. All his gear had been packed up. Stallard was gone.

As the rainy season wore on, it began to look as if the predicted fruit famine would not occur. The memory of the famine of 1971 was still strong among those who'd witnessed it. The event had reminded tropical biologists that rain forests, for all their complexity and diversity, are hardly as ecologically stable as scientists had once theorized. Floral and faunal populations soar and crash in a readily observable time scale. Fruit disappears and famine ensues. Trees topple, creating light gaps that kill shade-tolerant trees and give pioneer species hungry for light the chance to take off for the sky. The forest, it seems, is in perpetual motion.

But what sets this motion off? Apparently, it is the weather. Unlike the lowland Amazonian rain forest, BCI has sharply defined rainy and dry seasons. From December until the middle of April, the forest receives less than 400 millimeters of rain, and many trees lose all their leaves. From May through November, about 2,200 millimeters fall. This seasonal difference profoundly influences the biology of the island.

To describe and quantify the effects of this seasonality on BCI's forest, from the lowest trophic level to the highest, residents, starting in the 1970s, set out with their meter sticks and the latest in high-tech analytical equipment. They constructed elaborate sprinkler systems and watered plots of forest for five successive dry seasons. They deprived understory shrubs of water by laying plastic over the soil around their stems. They weighed leaf litter. They analyzed soil. They measured photosynthesis and carbon gain and water loss. They modeled the timing of leaf fall, leaf flush, and flowering.

To investigate the role of animals on this shape-shifting forest, researchers collected and counted pollinators, litter arthropods, and canopy pests. They imitated herbivory by punching holes in plants. They excluded vertebrate herbivores from plots. They estimated how many kilograms of fruit the island's fig trees produced, and how many of those kilograms were eaten by bats, by monkeys, and by birds. From small-island populations, they removed half the adult male spiny rats, then half the adult females. They counted the dead.

As the puzzle pieces came together, the forest assumed a new clarity. Scientists established, for example, that seed germination peaked at the beginning of the rainy season. And that dry-season rains, such as those of 1970, or an indistinct onset of the rainy season could disrupt flowering and fruiting, sometimes causing mass starvation among the forest mammals. In-vestigators found that flowering peaked near the beginning of the rainy season, coincident with a peak of insect abundance, when pollinators are presumably most numerous and active; that certain understory shrubs, by synchronizing their leaf production twice a year, created a temporary surfeit of new leaves, many of which survived because there were too few herbivores to eat them.

Dry-season food shortages, researchers found, limit popula-tions of spiny rats, agoutis, and pacas, and possibly many other species. When the fruit crop dips, agoutis dig up seeds buried long before, peccaries eat bark, pacas browse seedlings, coatis search for arthropods in the leaf litter. When the rains abate and the air becomes drier, anteaters give up ants in favor of ter-mites, which are juicier.

One study suggested that the onset of the dry season resets a biological clock in *Psychotria furcata*, telling the shrub when to produce new leaves. Another suggested that fruit-eating birds restrict their breeding during seasons of fruit shortage. Births of *Cebus* monkeys appear to be timed so that the infants can be weaned in October, when fruits they can reach, and comfort-ably handle, will be plentiful. *Artibeus jamaicensis*, the fruit-eat-ing bat, seems to do the same.

On the forest floor, plant-eating arthropods are most abun-dant during the rainy season. It is the lengthening of the day and increased atmospheric humidity that tells *Stenotarsus ro-tundus*, a fungus beetle, to begin regenerating the gonads and flight muscles it resorbed ten months earlier, during diapause. It flies off to mate just after the first heavy rains of the year.

It was stories like these that renewed my faith in scientists who mined narrow niches. If they appeared to sacrifice the big picture by, for example, tracking the growth rate of gonads on fungus beetles, this kind of work also opened up the possibility of real intricacy — of creating beautiful little puzzle pieces that fit into a shimmering whole.

*

People were always bursting into my lab, for one reason or another. First it was Bret with his bat, and then it was an entomologist looking for a butterfly net. One afternoon it was Guillermo Goldstein, sweeping through the door that connected his lab to mine. Guillermo had a round belly, wild white hair, and a white beard. He was Argentine and taught at the University of Hawaii.

Waving a branch about five centimeters in diameter and two-thirds of a meter long, he shouted, "Look at this! Look at this!"

"What?" I said. I was in the middle of something.

"Do you know what this is?"

I figured it had something to do with plant physiology, so my immediate reaction was not what it could have been. Guillermo didn't notice.

"Come check this out." He was both impish and avuncular, and so I walked around the counter that took up the middle of my lab. I generally tried to ignore what happened over on the far side; exhausted researchers were always dumping muddy backpacks on the floor there and using the battery recharger. An awful lot of other equipment never seemed to get touched.

Today, the members of Guillermo's team, a Harvard assistant professor named Missy Holbrook and a Mexican postdoc named José Luis Andrade, were busying themselves with plastic tubes, thermocouples, and batteries. Guillermo pointed out two needles stuck into his branch. He said they were measuring the volume of sap flow. Sap is mostly water, but it contains nutrients.

"The first needle is cold, the second one is heated by a thermocouple," Guillermo said. "The sap cools the second one, and we're measuring the difference in temperatures between the two, which tells us how much sap is flowing up the tree. Neat!" He smiled at me. "Do you understand?" I had the feeling he wouldn't go on until I did.

"I think so. It's the ratio you're interested in."

"Exactly." He beamed. A bucket of water sat on a small wooden table perched atop the counter. A length of tubing ran out of the bucket and down into a cuplike instrument that fit onto the end of their branch. At this moment the water was flowing through the branch and out into a plastic yogurt cup, one slow drip at a time. A piece of foam insulated the thermocouple and probes, and a piece of reflecting foil covered all of that. The probes could be set to take readings every twenty-five seconds, "or whatever," and were left on a tree for a week. To install the equipment on one tree took Guillermo's team two entire days.

Tonight, Guillermo was simply testing the calibration. He'd been using the method for three years, on trees near BCI's plateau, and had published a few papers on his results. He'd found that mature tropical trees take up between three and five hundred liters of water a day, and that ninety-nine percent of a forest's transpiration occurs in canopy trees. Something about this revelation bothered me.

"Plant physiologists didn't already know how much water trees transpire each day?"

"No!" Guillermo said, with exuberance. "We didn't know this. There was no way to measure it."

What these scientists did and didn't know continued to amaze me. Someone on the island knew the muscular output, in joules per second, of hovering hummingbirds, but there was no unanimous decision on how trees sucked water into their highest branches. Some ecophysiologists used a pressure bomb to measure water movement; others swore allegiance to a xylem pressure probe. When the inventors of the two devices were both alive, they wouldn't even speak to each other.

Guillermo went back to his lab while water filled the yogurt cup. In a couple of weeks, after this irrepressible teacher had gone back to Hawaii, I'd hear a report on the thermocouple measuring system from José Luis, at a Bambi. In the meantime, I was left to wonder if Guillermo's numbers might be useful to Robert Stallard, who walked so methodically through the woods, measuring so carefully where all the water went.

Early one Saturday morning, cued by the piercing whistle of *Las Cruces*, I ran down the gangway with loaded backpack and headed to Gamboa with Stefan Schnitzer and his field assistant, Brian Kurzel. Stefan, working on his Ph.D. at the University of Pittsburgh, needed help counting and measuring the lianas in a plot he'd marked out in the Cocole forest, near Rodman Air Force Base, just west of Panama City. The Pacific side of Panama was known as the dry side; the Atlantic, where he had another plot, was wet. BCI, on the rain gradient, fell in between.

Of his study, Stefan said, "This is mega. I can show why lianas are found in such high densities in tropical forests. I can show why there are more on BCI than near the Atlantic." His eyes gleamed. "One-quarter of all woody stems in a tropical forest are vines," he continued. "But no one includes them in diversity surveys." In all the meticulous work done on BCI's own fifty-hectare plot, no one had counted its lianas. One hectare of surveyed Panamanian rain forest contained 1,597 climbing lianas, which grew on forty-three percent of the canopy trees and could reach lengths of one thousand meters.

In the BCI jeep, we drove into the city and over the high arc of the Bridge of the Americas, which rises above the southern outlet of the canal. The bridge is the beginning of the end of the Pan-American Highway's Central American stretch. About a hundred miles to the east, in Panama's rough-and-tumble Darién Province — described by the consular section of the U.S. Embassy as "frequented by guerrillas, smugglers, and undocumented aliens" — the highway degenerates into a muddy track. Off to the left, in the Gulf of Panama, container ships and oil tankers line up, bows to the wind, awaiting their turn to traverse the hemisphere.

We hung a right on the far side of the bridge. Rodman looked deserted; the base had already been decommissioned.

Farther up this road, workers were blasting away at the rocky contours of Contractor's Hill, a high point on the western slope of the saddle through which the canal flowed. Culebra Cut, this narrowest part of the canal, was being widened so that two ships could pass at once.

Of all the engineering problems the building of the Panama Canal had presented, the incessant mudslides of Culebra were the most intractable. Culebra had defeated the French, and it added tens of millions of dollars to the Americans' expenses. The problem was substrate: the Cut was made up of dense, sticky clay layered over seventeen different rock formations, six major geological faults, and five major cores of volcanic rock. When rain saturated the earth, as it did on a monthly basis for half the year, the mud sloughed off.

The brown goo buried steam shovels and the railroad tracks, locomotives, and flatcars intended to move excavated material offsite. A major slide — hundreds of thousands of cubic meters of mud and rock was not uncommon — would take several months to clear. Throughout construction of the canal, engineers tried to stabilize the slopes by cutting them back, but they never found the angle of repose. As recently as 1974, a slide dumped one million cubic yards of earth into the canal.

It was quiet in the woods, and dry leaves crackled underfoot. The rustling reminded me of my first walk on BCI, down Armour Trail during the dry season. The forest here didn't impress me: it received less than 1,800 millimeters of rain a year (Panama's wet side, on the Caribbean, got 3,300 millimeters a year). I saw no mammals and heard few birds, though that didn't mean they were absent.

The plot was forty by forty meters, broken up into one-meter squares. Brian groveled: he sat, stood, squatted, and crawled along the forest floor to find, identify, and mark all the lianas with numbered metal tags. With calipers in hand, Stefan called out each liana's diameter at breast height, its species, and whether it made its living in the canopy, subcanopy, or understory. My job was to log the data in his Palm Pilot. It was tedious work, and the heat soon forced me to conduct my labors supine.

"Not much diversity here," I said, lazily typing in the symbol for *Doliocarpus* for about the hundredth time.

"Nope," Stefan said. "It's much higher on the wet side, which has a more distinct rainy and dry season, and even higher on BCI, where there's even more seasonality."

Working in his comparison plots, Stefan wanted to show that the more seasonal a forest — that is, the bigger its climatic range — the higher its diversity and density of lianas. Trees have the opposite scenario: their diversity and density diminish in forests with a great deal of seasonality. The lowland forests of the Amazon, which are rainy all year and thus have little seasonality, are biological hot spots for most plants, fungi, bacteria, and animals. But lianas, Stefan believed, do not follow this trend so closely.

In another study, Stefan was proposing that lianas can call the shots, in terms of what trees grow in gaps. "Gaps with high numbers of lianas suppress shade-tolerant trees," he said. "So liana-laden gaps result in a different successional trajectory than those without lianas." If so, then liana composition is a tool for predicting what a gap will look like in twenty or thirty years, and lianas can be a mechanism for maintaining diversity.

At the same time, however, vines hinder forest production. They depend on tree crowns for support, but by overtopping a crown, lianas restrict or prohibit its growth. Without a healthy crown, a large tree may not be able to produce a full crop of seeds or photosynthesize enough to maintain its health. Working on BCI, the vine expert Francis Putz wrote, "Scaling to the treetops may represent success for vines but it spells doom (or at least gloom) for trees."

Doliocarpus looked like nothing to me — a winding brown stem, its leaves nondescript. But as the rainy season progressed, the red fruits of *Doliocarpus olivaceous* and *Doliocarpus major* would abound on BCI, where they'd be devoured by the island's spider monkeys. Stefan would be gone by then: he'd never see his liana in fruit. Chrissy, in turn, knew nothing about this liana except that her spider monkeys ate its fruit; she wouldn't know that it grew faster and in greater numbers to the north than in the south.

In an age of niche specialization, where scientists move into ever-narrower fields in order to avoid competition and make names for themselves, it isn't unusual for a biologist to have deep knowledge of part of a study organism but be ignorant of the whole. A postdoc on BCI might be able to tell you anything you want to know about her butterfly's tympanic ear, but she knows very little about its foraging ecology. A professor who studies frog vocalizations understands the physics of frogs' throat sacs but not their reproductive biology.

Biologists, like other organisms, specialize out of necessity. Unlike the meter-stick science of a century past, today's science is expensive and dependent on public funds that have declined sharply as the number of grant applications has increased. Huge bureaucracies within the universities and funding agencies further diminish resources.

But fresh biologists keep coming. A top research university typically receives between one hundred and two hundred applications for every job in organismal biology. The increased competition for jobs forces graduate students to specialize earlier in an effort to carve out a niche that no one else has mined. The biological analogy is obvious: natural selection favors the competitor with an edge, and in a supercompetitive environment, specializing, at least in the short term, is a better strategy than generalizing.

Not that generalizing is much of an option anymore. Who has the time? Ten times as many scientific journals fill library shelves today than filled them fifty years ago; half a million scientists are at work around the world, doubling scientific knowledge every fifteen years. With so much information out there, it's nearly impossible for scientists to acquire the broad education that marked biologists of generations past. Only a handful of non-ornithologists can identify bird songs these days; a canopy specialist won't necessarily know his mushrooms.

Not only do today's scientists have less general knowledge, but they also tend to select research questions that are smaller and safer than those asked a generation ago. On the subject of grad students, E. O. Wilson wrote, "Most accepted low-risk projects, those new enough to generate significant results but close enough to preexisting knowledge and proven techniques to be practicable." A large part of "practicable" is "fundable."

Nearly forty years ago, a young theoretical ecologist from Princeton named Robert MacArthur came down to the tropics with his binoculars and the NSF's backing to look at the niches and food habits of warblers. His love of birds and his facility with abstract mathematics led him to his groundbreaking *Theory of Island Biogeography*, written with Wilson. In today's political climate, however, the NSF would probably tell MacArthur either to get lost or to produce for his warblers a phylogeny, an analysis of the historical relationships between different species that contains molecular as well as morphological information. (Phylogenies look like family trees, and their branches are called clades.) They'd ask him to spell out every methodological detail and expected result.

Sadly, unorthodox thinking and broad studies are now neither encouraged nor rewarded. Science is no longer the gentlemanly pursuit that it was when Charles Darwin slowly and elegantly worked out his theory of evolution as he strolled his country estate. Except perhaps for Egbert Leigh, slowly working out the significance of mutualism from his aerie near the laundry room, this approach is given little chance of succeeding in the modern era.

I found the counting and measuring in Cocole the epitome of drudge work, but I knew it was important: from raw data big ideas come. Wilson counted ants for years before his ideas coalesced in *The Theory of Island Biogeography.* Who knew where Stefan's rarefied line of inquiry would lead? His niche may have been narrow, but that didn't mean that wonderful surprises were out of his purview.

Punching in another series of numbers, I asked Stefan why he cared about the causes of diversity. Wouldn't his energy be better spent saving rain forests, because they were so diverse, than studying what made them that way? Stefan hadn't been at Lanting's slide show, but I had the sense he would have been moved to hear what had happened at Madagascar's Ankarana.

"If I wanted to save the rain forest I wouldn't be doing what I am," Stefan said. "I don't think of the big picture always; I don't want to. I'm setting up this experiment as an exercise in thinking. I don't want a utilitarian reason for everything. Why do we need art? I feel the same way about basic science: it's good for us."

We knocked off at four-thirty, crossed the Bridge of the Americas, and stopped for a beer at the Balboa Yacht Club, a wonderfully seedy marina, redolent of mold and vomit. Off the breezy veranda, sailboats bobbed at anchor. Prostitutes prowled the bar at night, hunting sailors waiting for ships. There was a board near the bathroom where people looking for line-handling work posted notices. Every boat that traversed the canal needed four people to secure the craft inside the locks — protection against the turbulence of nine million gallons of water pouring in and out. The going rate was sixty bucks a day, plus lunch.

We stayed that night at the Hotel Central, in the city's Casco Viejo, or old hull. The hotel was built in the colonial style, around an interior courtyard. During their attempt at the canal the French had used the Central as their headquarters. Younger BCI residents sometimes stayed here now — it was very cheap — but only if rooms were to be shared. Alone, the place could be scary; in company, it was merely funky.

Before heading out for dinner, we drank beer on our balcony and watched skinny dogs chase cats around the plaza below. Turkey vultures soared around the ancient Catholic church, just opposite, and black-clad *abuelas* shuffled inside. We played Paper, Scissors, Rock to pick who got which saggy bed and who got to shower first. For three beds we had paid a total of fifteen dollars. The hotel had no pillowcases. The floors were painted green, the transoms were open. One fluorescent fixture lit the room. The shower was a vertical pipe that trickled cold water into a clawfoot bathtub; there was no shower curtain. The water ran out a hole in the tub and meandered across the tiled floor to a drain in the center of the room. Our own private catchment.

I lay awake that night, listening to the cacophony of the street. It was stifling inside, though the windows were wide open. At around two in the morning another hotel guest tried to get into our room, scraping at our locked door with his key for what seemed like twenty minutes. Stefan and Brian slept through it. Eventually the stranger gave up, then put his key almost silently into the door next to ours.

Outside, a drunk yelled for hour after hour, berating some invisible antagonist. I heard the sound of retching. A heavy truck ground its gears as it monotonously circled the plaza. I wished the man and the truck would collide already, silencing both for good.

We rode *Las Cruces* back to BCI on Sunday. This was the weekend launch, the slow boat. She was cousin to the *Jacana*, named for the indigenous yellow-legged bird that seemed to hover over the water's surface as it took flight.

The old boat, though lumbering and plagued by chronic mechanical problems, had a lot more character. Fifty years old, *Las Cruces* had been bought from the Panama Canal Commission for ten dollars. Her upstairs deck was open to the sun and rain. Once a week residents requisitioned a "night boat," and *Las Cruces* would run from Gamboa to BCI at ten P.M., allowing residents to eat dinner in town and maybe catch a movie.

Sitting upstairs under an open sky was an incomparable delight. The air was warm, stars shone. We brought beer or wine along for the ride. Sometimes there was a guitar, sometimes a boom box and dancing. We followed the shipping channel, which carved its way over the deepest part of the old Chagres riverbed. Supertankers and the occasional cruise ship glided by, decorated, by nautical law, with strings of white, red and green lights. Layered and graceful, they looked like floating cakes.

Day and night, the canal was busy with PCC tugs and pilot boats, their hulls blue and cabins white. Gamboa was one of their operational centers: here dredging barges sat at anchor, shadowed by the giant cranes — named Olympus, Hercules, and Goliath — that lifted the steel gates from the locks. The gates weighed 690 tons, and they came out twice a year for maintenance.

Container ships from all around the world passed by: the *Industrial Beacon*, the *Murkur Star*, the *DMS Senator*, the *Baltic Mermaid*, the *Maertz Texas*. They plowed the water with their massive bows, creating wakes slow and thick. Their decks were five stories high. Flitting about them by daylight in our Boston Whalers — loaded with spiny-rat traps, madly chasing butterflies with waving nets — we were as gnats to giants. If they were curious about our exertions, we never found out.

While Brian and Stefan read on the upper deck, I contemplated the study of water. Both Guillermo Goldstein and Robert Stallard had systems for measuring it. For Stallard, all equations had to balance: he accounted for every millimeter of water that entered and exited a watershed. Guillermo's interest in water matched a tree's interest in water: how it got enough, converted it to sap, and moved it from earth to air. For Stefan, the rain gradient was at the heart of liana abundance.

Water flowed through the studies of three different researchers, but the researchers themselves did not mingle. Their puzzle pieces, as of today, shared no common edge.

Maybe the three men didn't overlap in time, on the island, or maybe it was space. In pools of black type, their data collected, to be published here and there. Someday, maybe Stefan would apply Guillermo's water-transport formulas to lianas. Or Guillermo would consider Stallard's water budgets when he measured sap flow in trees that grew near streams. Maybe something would come of it. Or maybe not. The boat glided peacefully through the jungle, and my tangled ideas about interconnecting puzzle pieces and isolated scientists drifted slowly apart.

9

Bug Jocks

URDER MYSTERIES and airport novelettes crammed the shelves of BCI's lounge, but now and then I caught a resident with her nose in a classic: Wallace's *The Malay Archipelago* or Spruce's *Notes of a Botanist on the Amazon and Andes*.

Like scientists propelled into polar research by the stories of Peary, Scott, and Amundsen, many of today's tropical biologists pore over the tales of the early naturalists. Some were inspired by stories from the more recent past, by Frank Chapman's *My Tropical Air Castle* or W. H. Hudson's *Green Mansions*.

In an age when so much of the world has already been explored, when the "big" discoveries have already been made, tropical forests are still supplying a frisson of adventure to those willing to consider the small scale. Amazonia may now be accessible to anyone with the proper resources and enough gumption, but the interactions of its forests are still terra incognita. Biology, in this sense, is one of the last great frontiers: it drives researchers to discover the creatures that make up our planet, to learn what these organisms are doing and how they do it.

For whatever reason, the world is still fascinated by the spirit of yesterday's scientific adventurers — the men and women who filled our museums of natural history — and by the hardships they endured for the sake of science and collection. Naturalists setting out for the tropics a century and a half ago needed to be mapmakers and anthropologists, sketch artists, general practitioners, marksmen, survivalists, and linguists. The English naturalist Richard Spruce, in addition to speaking Spanish and Portuguese, learned seven Indian languages. He was fluent as well in mycology, ornithology, primatology, and herpetology.

Barro Colorado's data collectors are the inheritors of this tradition. Their skills are almost entirely different, but they mine similar turf — in a much-changed world. Now, disease hardly factors, and physical hardships are minimal. Bates and Spruce sold specimens to finance their scientific exploration; BCI's researchers sell their ideas to the National Science Foundation, competing with scientists more numerous than the native Indians Bates encountered in the course of eleven years in the jungle.

Unable to go wider in this smaller world, today's researchers go deeper. Where von Humboldt had trouble simply identifying plants, now scientists examine their DNA, the way their cells give up water, and their relationship with the fungi that colonize their roots. A modern biologist can spend her entire career examining the relationship of one bee to its environment. Instead of looking at what an organism eats, where it lives, and how it reproduces — yesterday's brief — today's researcher might examine that organism's neurobiology, its genetic relationships and chemical arsenal.

I was loafing around my lab one night, playing with a tick I'd removed from my arm, when the back door hissed open. It was the pigtailed woman I'd seen up by the old dining hall for a couple of nights in a row. She seemed to be all business — not part of any clique and never appearing for meals. That was often the story with people who worked on nocturnal organisms.

When I first saw her, the woman had been shining a black-light onto a bedsheet hanging from the balcony. The UV rays attracted dozens of strange, cryptically colored moths. Some of them looked like dead leaves, others looked like bird scat, lace, or sticks. One of the showiest moths, a saturniid, sported giant eye spots on its nearly eighteen-centimeter wings. The insects in turn attracted nocturnal insectivores: bats, frogs, and the insect-eating *Bufo marinus*. The insectivores attracted their own class of researchers — mostly residents who'd knocked off for the day but couldn't resist messing around with the accidental menagerie.

Standing now in the dim corner of my lab, the braided woman said, "Oh, hi. Hi. Do you happen to have a syringe smaller than 1 cc? I'm trying to inject some solution into a butterfly's ear and what I have is way too big."

I was flattered she thought I was a scientist. "No, sorry," I said.

She began to scan the room somewhat urgently. "Do you think anyone else who uses this lab has one? Because I'm spilling solution well beyond the ear cavity with the one I've brought."

"This is basically a plant physiology lab. I've seen them stick syringes in tree limbs, but they're pretty big." I suggested she try the epiphyte people, upstairs. They did a lot of work with insects.

She thanked me and scurried out to continue her search, practically vibrating with anticipation. As the door swung shut I looked at my watch. That was an odd request for ten o'clock at night, I thought, smiling.

A few days later I was walking down the hall when the woman who had burst into my lab suddenly burst out of *her* lab. Her name, she said, was Jayne Yack. She was thirty-eight, a postdoc from Cornell working at Ottawa's Carleton University on a Canadian scholarship, studying the evolution of hearing in Lepidoptera.

"Come here," she said, pulling on my arm. "I want to show you my ear museum."

A Russell Stover chocolate box sat on the counter of her lab, surrounded by a dissecting microscope, a few cameras, black-lights, boxes of electrophysiology equipment for recording bat calls, a vacuum pump for sucking air bubbles out of moth ears so she could properly photograph them, and a dozen or so live moths clinging to the side of a black mesh bag.

Flipping open the lid of the box, Jayne revealed a half-dozen moth ears — they looked like tiny blobs of wax — pinned to pieces of Styrofoam. "This stuff is just so neat," she said. "You've got to see this membrane, the difference between a regular-frequency ear and a high-frequency ear."

She put two ears under the microscope and waited for my appraisal. The high-frequency ear had a much thinner membrane — it looked like yellow cellophane — and was backed by an air chamber. "You're looking at the chordotonal organ, the sensory cells," Jayne said. "It has three sensillae that stick up straight from the membrane. They come together in a nerve ganglion that sends signals to the brain, or what stands for the brain in the thorax."

Insects have ears on different parts of their bodies. Most moths have them on their abdomen or thorax. Crickets have ears on their legs, mantids on the ventral region of their thorax, cicadas on their abdomen. In Lepidoptera, ears are always close to the wing, because the ear cells need to communicate with the wings. "They need to command them in order to evade predators," Jayne explained. "The message doesn't go only to their head: they have a brain in their thorax and one in their abdomen too."

It had long been known that moths have ears, but it wasn't until the 1950s that Kenneth Roeder, a neuroethologist, discovered that their ears have between one and four nerve cells (human ears have more than fifteen thousand sound receptors each). He put his electrode on a nerve, played sounds, and watched the reaction in the neural pathways. In some species, he found, the ears were sensitive to high-frequency sounds. Why? In order to detect the echolocations of the moth's primary predator, the bat.

Other insects that have bat-detecting ears are lacewings, praying mantises, and also some beetles. There's an easy way to tell if a moth has sophisticated ears: "The jangling of keys produces ultrasound," Jayne said. "If the moth can hear it, it will dive — an avoidance behavior. It's a good way to impress your friends." Moths that can hear a bat approaching and respond to its attack, she found, have a forty percent lower chance of getting eaten than do moths insensitive to ultrasound.

On BCI, Jayne was looking at the ear of Hedylidae, the world's only night-flying butterflies, which for centuries were considered moths, since they make their living at night. Jayne had discovered during some routine anatomy that hedylids possess a high-frequency ear. "I'm trained to see ears," she explained.

She was studying the family's anatomy (that tiny syringe she needed the other night was for injecting a solution that would reveal the ear's anatomical structures), and she was documenting the behavior of butterflies, videotaping their swoops and accelerations as she sent out ultrasonic signals from a dog-training device. Next year she'd revisit BCI to make physiological recordings, hooking butterflies' nerves to electrodes to show their reaction to sound. With Elisabeth Kalko, the bat researcher, she'd study butterflies' reactions to bats, and find out if they produced their own sounds. There were forty species of Hedylidae in Central and South America, species about which science knew hardly anything.

Now Jayne segued from hedylids to tell me about a member of the family Sphingidae, which feeds on flower nectar. "This moth is crepuscular," she said, meaning that it feeds at dawn and at dusk. "And one of them, when it inflates its palps, its mouth parts, creates a tympanic ear." She gave me a look of triumph.

"Look at this one," she continued, turning to a small, dusty-looking moth hanging inside the mesh bag. "It's *Homeocera stictosoma*, a wax moth. It sends fluffy gray stuff that looks like dryer lint out from a vent in its abdomen when it's threatened." Little was known about the species, and Jayne had seen it only three times during her visits to BCI.

She brought the moth down the hallway to the epiphyte lab, where a German grad student named Christian had a video camera and a black scrim. Jayne held the moth in a pair of calipers while Christian taped. Riled, the moth performed as advertised. A thin stream of flocculent began to unspool from its belly. It floated upward — five centimeters, ten, fifteen. The stuff seemed evanescent, about to disintegrate. But when the fluff reached a third of a meter, the moth had a change of heart and began to reel it back in, scrambling with all six legs to retuck it in the cavity from which it came, to be used another day.

A look of satisfaction settled upon Jayne's face. "I'm sending the fluff to a guy who does chemical ecology," she said. I asked her where she'd found the moth. "I was on Pipeline Road, in Gamboa," she said. "I knew that many moths taste bad and smell bad, to discourage predators, so I just held it right up to my nose and gave it a good sniff. And this stuff just shot right in!" She guffawed, then gently placed the *Homeocera* back inside her mesh bag.

As we walked back to her lab, Jayne explained that looking at lepidopteran anatomy and behavior gave scientists clues to the evolutionary origin of this insect order. (Studying how the moth's sensory structures governed its behavior also gave scientists insight into how the human nervous system works.) Fossil records show that butterflies, which evolved from moths, became daytime flyers at about the time, millions of years ago, that bats developed large ears and the ability to use echolocation in hunting. Moths, too, developed ears at about this time. Jayne believed that the common ancestor of butterflies and moths was atympanate, or nonhearing: neither family needed ears until bats that could echolocate appeared on the scene.

Leaning against her lab counter, she said to me, "Either Hedylidae was a diurnal atympanal butterfly group and it moved to the night and developed an ear, or it comes from a butterfly ancestor with ears, but the rest moved to the daytime and lost the ear." This all came out in a rush. Now Jayne took a breath. "This is so much fun!" she said.

After returning to Canada, she would conclude that the second of her hypotheses was correct: that hedylids are the more ancient butterfly, that they kept their ears because they never moved to the daytime, the way other bat-evading butterflies did. The butterfly, she'd later write, was, in effect, "invented" by the bat.

I liked Jayne from the moment I met her. She was overtly excited about her work, and she wasn't reluctant to share her enthusiasm with someone far outside her field. Younger residents, I was finding, could be somewhat more circumspect. There was the parasitization factor, of course. But younger scientists also had less knowledge to share. Perhaps competition and the pressure to specialize kept them from poking lazily around the forest, from developing enthusiasms for organisms beyond their vocational scope.

Or maybe I liked Jayne because Jayne liked insects. And insect people were just about the most-fun scientists to be around. Years before, E. O. Wilson had told me as much, and my months on BCI were beginning to confirm it.

In *Bright Paradise*, David Raby wrote that Henry Walter Bates was, "because of his need to peer closely in search of his favourite insects, brought into close quarters with every kind of living thing." Here on BCI, an entomology grad student, in a fit of chauvinism, told me, "Ant people are broader than other biologists because ants figure in so many things — soil regeneration, plant death, parasitic relationships." Steven Pinker, director of the Center for Cognitive Neuroscience at

MIT, has called entomology "the ultimate multicultural curriculum."

It wasn't hard to run into bug jocks on BCI because the island, like most tropical rain forests, contains vast numbers of bugs. (Strictly speaking, a bug is a member of the order Hemiptera, insects with modified forewings, and the word should be used only when true bugs are the subject.) There was no accurate arthropod count for BCI, but taxonomists have formally described a million different insect species, the majority of which live only in the tropics. Experts believe there may be as many as thirty million more tropical insect species waiting to be described.

Sabine Stuntz, with whom I'd have several hair-raising nautical adventures in the months to come, was trapping and counting a portion of those insects. Sabine hypothesized that epiphytes contribute significantly to the canopy's diversity of arthropods (a phylum that contains insects, arachnids, crustaceans, and myriapods, or centipede-like animals) by providing food for herbivores, creating niches where they can live and possibly mitigating the harsh and arid conditions of the upper canopy. Like so many of the insect world's niche specialists, Sabine focused her energy narrowly — in her case, on the creatures who inhabited just one species of tree.

The tropics have long tantalized scientific thinkers with their luxurious diversity and with hints of Earth's primordial conditions. Darwin called the tropical forest "primeval," a place where "the powers of Life are predominant." In 1922, David Fairchild, writing for *National Geographic*, called the tropics "the mother liquor, so to speak, from which the plants of the world have come." In Panama, he felt as if he were in the world as it was before man existed, where "life teems and new forms develop, in the midst of that living stuff up out of which man came ages ago." In *Naturalist*, E. O. Wilson wrote, "The tropics I nurtured in my heart were the untamed centers of Creation."

The tropics have been in business since at least 3.5 billion years ago, when the first photosynthesizers figured out how to use sunlight for energy and began releasing carbon and oxygen into the atmosphere. As recently as the Eocene period, 38 million years ago, tropical vegetation extended all the way up to southern Alaska and Labrador. It has been reported that a majority of temperate lineages are evolutionary derivatives of tropical precursors. The temperate zone, then, is an aberration, and the tropics are the norm against which the farther latitudes are measured.

But why, in that case, is so little known about the tropics? How can there be several million insect species — to name just one group of organisms — still unknown and undescribed?

In short, because scientists have been studying the tropics for less than two hundred years. And why is that? Because the tropics have long been considered an uncomfortable and inconvenient place to work, and because most scientists have a temperate bias, and because funding to study the tropics is not abundant. The countries that contain the tropics have little money to spare for basic research. Granting agencies in the United States prefer to fund studies within U.S. borders, and most money, even for studying such global phenomena as El Niño, ends up in the temperate zone.

In a joint paper, Bert Leigh and Ira Rubinoff, STRI's director, pointed out that enormous advances in tropical biology could be made at a fraction of the cost of a single space launch. If funds matching those of the Human Genome Project were devoted to tropical research, they wrote, "we might find ourselves in a better environment in which to employ our new understanding of human genetics."

Science's understanding of tropical nature was in its infancy, and my own understanding of the field was barely gestational. But as the months passed, I felt as if I'd made some slight progress, if only because I could now identify a few more plants.

By its thick pink flowers, I recognized *Gustavia*; by the

orange casing over its red-ariled fruit, I knew *Virola*. By its peltate leaves I could pick out *Cecropia;* by its waxy leaves, *Anacardium*. The spiky oval seedpods belonged to *Pithecoctineum cruciferon*, the heart-shaped leaves to *Platypodium elegans*. *Pseudobombax* was the tree with shaving-brush flowers; *Socratea* had prop roots. If I came upon a round green fruit the size of a plunker marble, I knew *Callophylum longifolia* was not far off.

I was in love with the scientific names of plants. They were poetry to me. Why did knowing a few names make me so happy? Because it meant that I was starting to classify. As Adrian Forsyth and Ken Miyata wrote in *Tropical Nature*, "Complexity excites the mind and the discovery of pattern rewards it."

Unfortunately, I soon realized that patterns would get me only so far. If the tree lacked distinctive leaves, if there were no fruits or flowers, I was lost. I knew monkey comb's seeds but not its bark. I knew *Astrocaryum*'s bark but not its leaf. It all seemed frustratingly complex to me, at times even unknowable. I kept bumping into trees that looked alike but were different, and trees that looked different but were alike. All of them, I learned, had ingeniously different solutions to such common problems as shedding water from leaves, dispersing their seeds, and defending themselves against herbivores.

Some trees flushed their leaves in pulses throughout the year, some flushed them continually, some seasonally. Other trees were deciduous and lost all their leaves for a regular period each year, or only at the advent of the dry season. Seeds ranged widely in their size, weight, architecture, and seasonality of dispersal. Each tree species attracted different pollinators and dispersers. So many different strategies!

The odds of any one tree growing to adulthood were so long that biologists referred to the startup system as a lottery. Seeds could disperse from the canopy but get hung up on branches and never sprout. They might drift on winged pods, called samaras, into Gatun Lake and drown.

If a seed made it to the forest floor, it had to escape predators and pathogens, germinate successfully, survive the predators and pathogens of its youth, and suppress its growth in the shade of the canopy but continue to build its reserves until a patch of light penetrated the forest's upper strata, thanks to a treefall, a broken branch, or even the loss of a single big leaf. Only when light hit the seedling would it be released to grow into a canopy tree. For some, this enforced adolescence lasted decades.

A lot of things had to go right for a tree to survive the forest lottery, and they didn't go right for all species at the same time. Obviously, "enough" trees made it to adulthood. But which tree lineages would persist over the long term, the real measure of success? How did one weight the forces of speciation, immigration, extinction, drift, predators, pathogens, and a myriad of abiotic disturbances?

Sometimes I felt intrigued by the forest's complexity. Sometimes I was overwhelmed. And sometimes I found the forest . . . monotonous. This feeling was reactionary, I knew; it was the response of the benighted — or perhaps the demented, to use von Humboldt's expression. I lacked the training to recognize either individuals or patterns. I wasn't making the crucial connections. I was tired, perhaps, of thinking.

Tropical forests, says Thomas Lovejoy, of the World Wildlife Fund, are a world for biological sophisticates, a world whose wonders become apparent only with considerable patience and background. Indeed, most first-time visitors to the tropics are disappointed: the animals keep to themselves, and the greenness seems unending.

Learning about different ways to approach the forest didn't always advance my understanding, but I kept going out with scientists anyway, pitching in when I could, watching and listening when my hands were not needed. Bert continued to identify trees for me. Sometimes the names stuck, sometimes they didn't.

What I remembered best from my guides were stories about interactions, both competitive and cooperative, between organisms. They were narratives that I could easily grasp. For example: A parasitic wasp, studied in Costa Rica, temporarily paralyzes a spider until she can lay an egg on the arachnid's abdomen. The spider returns to building her daily orb web until the wasp larva, ready to kill its host, chemically induces the spider to build a cable-supported platform, safe from predators below. When the larva's new home is complete, the spider dies.

There was no end to such tales because the forest contained a vast number of twisted relationships. It was just what you'd expect, really, from a community that contained a vast number of idiosyncratic creatures.

The complicated relationships of mutualists, parasites, and commensalists (in which one species gains but has no effect on the other) went far to explain how so many creatures were accommodated and sustained. But this left, of course, the million-dollar question: What made so many creatures arise here in the first place? What explained the tropical riot?

The latitudinal diversity gradient poses the question this way: Why does species diversity go up as latitude declines? Besides its theoretical import, the question has strong practical implications. How can conservationists preserve tree diversity if they don't know what causes it? (Because the diversity of plants is the foundation for the diversity of plant consumers, most diversity-gradient theories focus on why there are so many kinds of tropical trees.)

I was inclined to tiptoe around this subject. It confounded me, and it confounded, I suspect, many scientists. Even Darwin, in his *Origin of Species*, never really explained how species emerged in the first place.

Over the years, theorists have devised numerous arguments to explain the diversity gradient, arguing for the power of stability, time, productivity, predation, specialization, competition, and numerous other biotic and abiotic factors. Many of

these theories contradict one another, many rest on circum-
stantial evidence, and many have strong historical components
that make them difficult to test. As Bert Leigh wrote in *The
Ecology of a Tropical Forest*, there is no "unified and harmonized
body of dogma" about the tropical forest.

Studying John C. Kricher's *A Neotropical Companion* late at
night, I read about the stability-time hypothesis. This theory
holds that the moist tropics are so old, and have been so climatically stable (escaping glaciation, among other extremes),
that large numbers of species have had time to speciate here.
Fewer environmental stresses mean less extinction, and so organisms "pack" into the tropics over time.

In the absence of a perfect fossil record, this idea is not testable. Moreover, a forest as seasonal as BCI's does not offer a
year-round superabundance of food to its animals. The forest,
as Robin Foster learned, has seasons of unpredictable depri-
vation.

Forsyth and Miyata, authors of *Tropical Nature*, discount the
effect of age and predictability and suggest that tropical diversity instead derives from "the constant change and disturbance
engendered by falling trees." Year-round, but especially during
the rainy season, rain forest trees are blown over by wind and
pulled down by vines and lianas that loop for thousands of meters between trees. The shifting mosaic of light produced by
such treefalls presents opportunities to a myriad of organisms
who will also, of course, face shifting competitive pressures as
the gaps fill in with greenery.

A mechanism that saves trees from falling may itself contribute to diversity. According to one theory, trees that evolve different architectures and flexibilities can sway out of sync with
their neighbors and snap their lethal vine connections. In this
way, lianas, by pushing trees to differentiate, may promote an
increase in the diversity of rain forest trees.

Kricher presents not only the stability-age hypothesis but
also the interspecific competition hypothesis. This theory argues that competition among species for limited resources —

such as light or nutrients — results in increased specialization. These highly specialized species pack into extremely narrow ecological niches, which reduces competition among them for their exclusive resources.

Just opposite to this idea is the predation hypothesis, which holds that predators, which continually switch their attention to the most abundant prey, preserve diversity by preventing one competitor from pushing another to extinction. In this scenario, specialization is less likely to occur because predators keep competition levels low.

The intermediate disturbance hypothesis argues that the tropics' moderate levels of abiotic disturbance (as opposed to the high levels in, say, a temperate mountain pass) reset the competitive clock with enough regularity to prevent extinctions. Competitors are always in the process of excluding each other, the theory acknowledges, but in the tropics, disturbance repeatedly interrupts the process before it reaches its conclusion. And so organisms accumulate from immigration and speciation.

The refugia hypothesis suggests that the glacial advance of the Ice Age, which began about two million years ago, made the Neotropics cooler and drier. Grassland savannas expanded, leaving behind islands, or refuges, of tropical forest. Geographically isolated from other populations and subject to subtle shifts in temperature and rainfall, species within these refuges, which were not accustomed to a variable environment, speciated widely. New life forms originated. "The importance of the refugia hypothesis," writes Kricher, "is that it attributes a significant part of the high species richness of the tropics to the effects of *instability* rather than the older concept of the stable undisturbed tropics."

Investigations conducted around the world and on BCI have supported or disproved parts of many of these theories. Studying data on water and nutrients on the island's fifty-hectare plot, Robin Foster and Stephen Hubbell concluded that it wasn't small-scale adaptations to a multitude of niches (in other words, niche specialization) that put three hundred species of trees on the island's plateau. There just weren't that many niches, they found. Rather, it was the race to get established first and to make the best use of space, nutrients, and microclimate. For Foster and Hubbell, the ability of trees to respond so differently to minor environmental fluctuations and other perturbations is crucial for so many plant species to coexist in such close proximity.

Their "lottery-competition" model recognizes that the more similar the competitive strengths of the species limited by a given resource, the longer it will take for them to outcompete one another. They divide tropical tree communities into a few guilds populated by species close in competitive strength, and they predict that species densities within a guild will go up and down randomly. In this scenario, competition between species has no structuring effect on community composition.

"There are likely to be elements of truth in each of the diversity hypotheses," writes Kricher. "Perhaps that is the real key to tropical diversity. Several forces are working simultaneously in promoting and preserving species diversity."

On that note, I let the matter rest. There were other, more immediate questions that scientists on BCI were addressing. But the diversity question did not go away, and before long I'd be forced to consider it once again.

10

Random Error

I MET BRET one afternoon on the balcony of the old dining hall to talk about his research and sip a gin and tonic. From here, we had a terrific view of the lake and its passing ships. Gloria, a two-and-a-half-meter-long crocodile, patrolled near the raft. High above Laboratory Cove a snail kite rode a thermal. These birds of prey, endangered in the Everglades but common as agoutis on BCI, specialize on a large marsh snail called *Pomacea*, picking them off the dead trees that poke from the lake's surface and prying them open with their hooked bill.

"I don't know how I feel about my project now," Bret said, shrugging. His tone was conversational. His feet, in hiking boots, rested on the varnished railing. "My data aren't great, and I may not be capable of getting great data."

The problem was the kernel method, a system used for estimating home ranges. The computer program took Bret's triangulation points, which were supposed to be recorded every six minutes, then mapped them into a cluster of dots, called a kernel, that represented the area in which a bat had been active.

On a sheet of legal paper Bret drew a tight grouping of dots and some dots farther afield. "The computer says here's your ninety percent kernel." He circled most of the dots. "This is where the bat roosts ninety percent of the time." Then he enlarged the circle to include some outlyers. "This would be your ninety-five percent kernel."

It was the same for flying: the computer spit out a row of dots that showed, roughly, the bat's line of flight. "When you track a bat that's making long flights, you'll get some erroneous numbers from when it's close and better numbers from long-distance readings," Bret explained. "The closer the bat to the triangulators, the more error you get, because the bearing can move 120 degrees in a matter of seconds and the triangulators can't make instant readings. If the animal is far away, the observers' angle on it doesn't change much from moment to moment, and the bearings don't need to be simultaneous.

"Statistics allow you to correct for a certain amount of error," Bret continued. "That's what the kernel method was designed to accomplish. But when the animals are close, the error becomes large and the noise in the data tends to overwhelm the signal."

Watsoni and *phaeotis*, Bret had discovered, don't fly for very long: between five and eight minutes. And it was taking us anywhere from fifteen to ninety seconds, and sometimes longer, to get our bearings: the thick vegetation, hilly terrain, and changes in humidity and temperature bounced the signal around. Still, Bret knew the bat wasn't going far: the transmitter broadcast only three hundred meters, and we never lost the signal entirely.

"I'm calling what you get on the computer map an artificial scatter," Bret said. "The kernel looks tight, as if the bat is foraging within a certain area, but what it's really telling you is how much error is in the data."

Science recognizes two kinds of error: systematic and random. Systematic error works in a particular direction and can therefore lead to erroneous conclusions. Random error just makes it hard to draw any conclusions.

"It's funny," Bret said. "Although the source of the error in this case is noise, the effect is systematic rather than random. The noise results in an overestimate of the bat's total range, suggesting that the animal was tracked in places it never actually went."

He paused for a moment. "At least the direction of the systematic error is good," he said. "If my data show small ranges in spite of the tendency of this method to overestimate, then it means the ranges are really small."

He went on, with a grim smile. "I think I've got a pretty good idea about what these bats are doing, but a lot of that is based on anecdote. I need real numbers." To prove that his bats were saving on commuting costs, he needed to compare their average home ranges with those of another small frugivore that didn't make tents. Such was the scientific method. Without a comparison, his prediction would remain untested.

BCI had few small, non-tent-making fruitbats, so Bret would compare his bats' ranges with those of *Artibeus jamaicensis*, a larger frugivore, and then correct for body size. The method was imperfect, but the level of uncertainty didn't bother him unduly. He wasn't running an experiment, he was looking at a historical pattern in evolution. And history, quite often, just doesn't give that much to go on.

We sipped our drinks. Sounding frustrated, Bret said, "I'm perfectly happy to go out every day and mess around, look for new tents, photograph bats, use infrared cameras, note a new behavior. I'm just not sure I want to pay the price I need to pay in order to get a degree."

That price, of course, was the drudgery of regular data collection. Bret sighed. "I could bail now and get a master's."

The truth was that Bret was more interested in what his animals were doing than he was in earning a degree. In this sense he was a maverick, bucking the tide of striving doctoral candidates. Making good over the field season drove the majority of BCI's younger residents. Their time here was limited, but the pressure went beyond collecting data industriously and in a clever manner. Researchers also had to publish, cultivate the people who mattered politically, and work themselves into the fabric of the scientific literati.

But none of that, apparently, mattered to Bret. He hadn't been in touch with his university for a year, and because he'd neglected to call a travel agent in time, he was shut out of a bat conference in Brasilia.

What Bret wanted, he told me, was freedom: to follow his instincts, to think, and to do good science — not just science that was safe and fundable. He said, "Scientists are supposed to be attempting to prove theories wrong, falsifying hypotheses, not just reaffirming others' work."

But would Bret's brand of thinking, in late 2000, convince his thesis committee that he could make a contribution to science? And was the contribution that he wanted to make the sort that science wanted? Bret raised his eyebrows and shrugged. "They may want someone who's more focused, not as far-ranging."

Still getting lukewarm results from radiotracking, Bret turned his attention to videotaping. During the first month of his field season, he'd set up his infrared camera, the kind used in bank surveillance, under a modified heliconia leaf. Near it, with a second tripod, he positioned a Bat Brite 4000.

The Bat Brite wasn't something you'd find in a zoology supply catalog, or even a photography catalog. Bret had invented and constructed it on BCI, using stuff he bought in Panama City. The Bat Brite consisted of 247 LEDs from TV remotes — they give off infrared light — hand-wired onto a circuit board and embedded in a paperback-sized block of Lucite. It wasn't elegant, but the Bat Brite was waterproof and could run all night off a five-pound rechargeable battery.

The first tape Bret made with the Bat Brite set a remarkably high standard for videos to come. Inside the tent an *Artibeus watsoni* nursed her young. She sucked the juice out of a fig that was almost as large as her body. She groomed her baby's wings, biting out tiny parasites. Then something bumped into the tent; the mother's nose and ears began to twitch. A male flew in and mounted her. The female bat screamed — on the video all her teeth were bared. She fought. The male finished and flew out. Bret had seen ten to fifteen bat copulations in his lifetime but never a rape. Neither had any other bat scientist, as far as Bret knew.

In tapes made over subsequent months, Bret captured a male licking its tent roof — a behavior never before described — and another constructing a tent in a heliconia. To build the tent, the bat used its flight muscles to tear the leaf, straining to rip the thick fibers with the thumb-claw on the leading edge of its wing. "The literature says they do it with their teeth," Bret noted. He was the first person, ever, to observe tent making. He had done it in the wild, and he had captured the entire thing on videotape.

One night Bret and I spent several hours setting up sound and video equipment under two partially modified philodendron leaves in Allee Creek. One leaf contained a bat, the other didn't. Bret wanted to see who was making tents, and how.

I left the site early, but Bret stayed in the creekbed until six A.M. He passed the time repairing a short in a cable and, when his rock perch became too uncomfortable, collecting all the bioluminescent wood and leaves he could find. He made a small pile that was visible, with the dark-adapted eye, from a few meters away. What made the fungus glow, and why? Was the glowing adaptive — did it help the organism survive — or was it the byproduct of some other process? Did the fungus need sunlight to glow? Eventually Bret decided that he'd need a solid background in chemistry to find out — and he never had cared much for chemistry.

Bats flew in and out of both leaves that night, but neither tent underwent any improvement. After he'd invested fourteen hours of labor on this night, Bret's N — the number of times he'd filmed tent making — still equaled 1.

Bret knew that collecting field data on bats would be tough, but he didn't mind the physical labor or the long nights. His problem was staying focused. Impulsive and creative, Bret tended to follow his wandering mind.

In a workshed, he was grinding a new bridge for the island's guitar. He was photographing a Jesus lizard — so called because it walks on water — and a baby crocodile he'd hand-raised and a sloth that had been hanging out in Lutz Ravine.

I bumped into Bret one afternoon near the old lab clearing and he pulled me toward a meter-long rolled-up heliconia leaf near ZMA dorm. "Look," he said, gently bending the foliage my way. Inside were two adult bats and one baby, their legs bridging the diameter of their green tube. Bret carefully reached in and removed an adult. Its wrists and ankles had curious pink disks that stuck to my fingernails like adhesive. The sucker-footed bat, *Thyroptera tricolor*, eats insects and roosts inside furled heliconia and banana leaves, clinging to their smooth walls with their suction cups. Because their homes unroll after a day or so, tricolors must constantly move. They were not rare on BCI, but they were difficult to net.

Bret opened the bat's mouth to show me its pointy white teeth. Unlike some people on the island, Bret wasn't reluctant — in the philosophical sense, not the physical — to handle animals, to stick a flash in a bat's face, to poke around in burrows until the inhabitant came out. He had none of the leave-no-trace leanings of, say, Hubert Herz, who carefully replaced any tree limbs he moved from his path. Bret wouldn't stress an animal unduly, but still, a mantle of entitlement seemed to settle upon him as soon as he picked up a camera.

This sense of entitlement was not new to naturalists. In the Galápagos, Darwin knocked hawks off tree branches with the muzzle of his gun, testing their fear of man. He rode atop tortoises and repeatedly threw a marine lizard as far as he could into the sea. Turning his attention to the lizard's terrestrial cousin, he pulled its tail as it burrowed into a hole and then harassed it with a stick, just to see if it would bite. It did, "very severely."

On BCI, most scientists felt free to muck around with animals without the public opprobrium they might feel back home. American popular culture didn't always paint scientists benignly, but here on BCI such prejudice was irrelevant. If the scientist was a cold-hearted nerd with a Palm Pilot, it hardly mattered: at least he had the support of his cohort.

I couldn't help worrying about Bret. Instead of working on his dissertation or radiotracking more bats, he was hunting mushrooms on a mainland peninsula and larking around on canopy ropes. When he told me he was experimenting with Saint John's wort to focus his concentration, I was glad. He took it orally for a few weeks and then tried smoking it, for speedier results. It was difficult to light, though, and soon he gave it up.

Bret might have been better off studying bats on BCI thirty years earlier, when the station was guided by Martin Moynihan. The irascible animal behaviorist valued keen observation and sharp insight far more than statistical rigor. His philosophy was simple: a researcher could save a lot of time by being intelligent; there was little point in collecting reams of data.

I checked in with Bert one night, to see what he made of Bret's behavior. Over a Bach cantata and a bourbon, he pronounced, "Bret seems to spend a lot of time doing nothing, but he's thinking. When he does act, it will be important."

Lately, Bret had been thinking about a way to escape the constraints of his radiotracking equipment, about a better method for capturing the bats roosting under Smith House, and about the latitudinal diversity gradient, my intellectual bête noir. Bret had his own spin on the question: Was the fact that species diversity was greater in the tropics tied to the fact that tropical species tended to be specialists, as opposed to the generalists of the temperate region? An oak tree, he explained to me, must allocate its resources to survive a broad range of climatic conditions. But a tropical tree can put its energy into specializing, the better to beat out the local competition.

The generalist-specialist question began to obsess Bret. He tossed his ideas around with visiting scientists, and he searched the Internet for references to support his hypothesis. Soon, using tradeoff theory to address species diversity would subsume Bret's interest in bats. When he returned to his university several months hence, his unsuspecting advisor in the department of evolutionary biology would learn that Bret's dissertation would be based, for the most part, not on field data but on theory.

Island scientists were constantly asking themselves why organisms look the way they do, asking what forces shaped their morphology and their behavior. In contemplating why the tropics have more diversity than the temperate regions, then, one is incessantly drawn toward the contemplation of evolution itself.

In his lifetime, Charles Robert Darwin made but one long voyage, aboard the H.M.S. *Beagle*, from 1831 until 1836. He traveled little after his return, but he continued to compare specimens of plants and animals from around the world. The apparent relationships between organisms separated widely, both by space and by time, told Darwin that new species arose through "descent with modification" over time.

In his masterwork, *On the Origin of Species*, published in 1859, Darwin stated that the mechanism for modification of species — in other words, evolution — is natural selection, wherein only the fittest of individuals survive to pass on the traits or behaviors that confer an advantage.

Over millennia, new species appeared as organisms adapted to changing environmental conditions. Even complex organs and instincts, Darwin wrote, arise through "the accumulation of innumerable slight variations, each good for the individual possessor." The essential prerequisite, he stressed, was time — lots and lots of it.

The longer I spent on BCI, the more evidence of evolution I

saw — in the range of biodiversity and in the endless variations on similar forms. But it was difficult to comprehend the sweep of time here. Unlike my temperate haunts, the island had no sweeping vistas of granite, no eroded canyons of sedimentary rock. Nearly everything in the forest was vegetated, and time seemed to pass at warp speed. Plants grew quickly, then decayed and rapidly transubstantiated into newly living matter.

The Isthmus of Panama was relatively young — a mere three million years old — but the lineages of the creatures here today were ancient. If one imagines the history of life on Earth, some 3.8 billion years, as taking a single year — a metaphor borrowed from Sir David Attenborough — then one day represents ten million years. In this construct, algae-like organisms don't appear till the second week of August. Worms start to burrow through the mud in the second week of November, and the first fish with backbones appear in the limestone seas a week later. Lizards — their legs derived from fish fins — scurry across beaches in mid-December, and humans appear on the evening of December 31.

Darwin, through his experience with breeding pigeons, had some understanding of heritability, but he had no knowledge of genes, which provide the variation upon which natural selection can act. Sexual reproduction, the commingling of genes, provides the most immediate source of variation. But even greater potential change comes from random mutation of genes.

Genetic mutations — the result of cosmic radiation, ultraviolet light, and copying errors — produce variety in two ways. Either the building blocks of proteins themselves can produce something new or, more commonly, the instructions for assembling those building blocks can change. Such changes are called regulatory mutations, and they're expressed during embryonic development, as cells multiply and proteins assemble in a nascent organism. The biogenetic law of Ernst Haeckel, a nineteenth-century German natural philosopher, holds that the embryonic development of the individual retraces the evolutionary steps of its ancestral group. Or, more pithily put, "Ontogeny recapitulates phylogeny."

In humans, Haeckel's law is seen in action as embryos pass through stages reminiscent of fish, amphibians, and reptiles. At some point in their embryonic development, all terrestrial vertebrates, today and for the last three hundred million years, sported gill arches, just like their marine ancestors. A mammalian heart develops through fish, amphibian, and reptilian stages until it becomes a four-chambered organ.

Understanding the developmental process is important because new species arise through both speeding it up and slowing it down. In some species, what was an adult stage in the ancestral species becomes a mere phase through which the embryo of the new species passes. Eventually a novel adult form is reached. The opposite happens as well. Numerous features of the mature human — including a bulbous cranium, vertical face, and small brow ridges — are typical of embryonic but not of adult apes and monkeys. Examples of retardation of development are numerous in the living world, evidence that altered timing through regulatory mutations can have a profound effect on the development of new species.

Darwin stressed that random and accidental changes don't stick unless they confer an advantage. Only "profitable deviation of structure or instinct" (including, as we'd come to learn, either selfish or altruistic behavior) is preserved. If mutations are large enough and beneficial, new species might eventually result.

Environmental challenges — geographic or climatic change — are the gauntlet through which mutations must pass. As conditions on Earth changed — as glaciers advanced or mountain ranges thrust upward and forests gave way to grasslands — individuals more adaptively fit for survival became reproductively isolated from their mother populations. The process of individuals spreading into new niches is known as adaptive radiation.

The release of competitive constraints can also spur radiation. Small mammals, freed by the Cretaceous extinction of dinosaurs sixty-five million years ago, moved from their nocturnal niche into the daytime, where they exploited herbivore and carnivore niches. New species may coexist side by side in slightly different niches, or, responding to new selective pressures, a new species may compete with and replace an older one. Rapidly changing patterns of mate preference — sexual selection — may also result in reproductive isolation and species formation.

In the tropics, I learned, abiotic conditions have remained relatively stable, compared to conditions of the temperate zone. Plant and animal species have accumulated — owing perhaps to low extinction rates — and worked their way into one another's lives, tighter and tighter, until many of them have become interdependent.

While the temperate zone doesn't lack for fancy adaptations, it's obvious that the tropics contain a lot more of them. As Stephen Jay Gould has said, "Increasing complexity of design begets more opportunity for variation upon it." In other words, complexity leads to more complexity.

But how did complexity get such a roaring start in the lower latitudes? A century and a half after Darwin and other scientific thinkers worked out the mechanisms of evolution, we're still trying to fathom why the tropics contain so much more of it.

I went home to marry my fiancé, Peter, and to forget about tropical abundance for a while. By the time I got back to BCI, the lake level had finally begun to rise. I had a roommate now, a woman doing a pilot study of coatimundi vocalizations. I met her at dinner, where there were plenty of new faces. I realized, taking a look around, that I was no longer a stranger here. I had come back; these people were just starting out.

Chrissy greeted me with a kiss. Bret, as usual, was absent. Paul Trebe was full of news. His six-month tour was almost up, and he looked like a new man. He'd lost thirty pounds since he arrived, he'd shaved his head, broken up with his girlfriend in Wisconsin, and started dating a Panamanian botanist. He'd told me several months before that he was looking forward to dry midwestern air, pine trees, and decent beer, but now he was signing on for another stint in the field.

Greg Adler, Paul's boss, had devised a new drama of population demographics in which his rats would star. In his last manipulation he'd added food to his islands; this time he was subtracting rats.

"Female spiny rats have two or three pups per litter, up to four times a year," Paul explained to me. We were sitting in the bar now, my first night back. Rat populations go up and down. But what regulates the litter size and its frequency? What are the limiting factors?

For spiny rats, as for any animal species, it's the individuals and not the population that regulate reproductive output based on the tolerances of the group. Was it the males or the females who were calling the shots here? Adler suspected that male behavior in some way limited reproduction. Adult male rats are territorial. They interact with juveniles and with sub-adults, excluding them from burrows and aggressively defending their own turf. If population change is density dependent, Adler theorized, then per capita reproductive effort should diminish as a population grows.

To test his hypothesis experimentally, Adler first boosted rat populations through provisioning, and then on four islands he ran his traplines and collected half the adult males. He moved them to a temporary colony off-island. Four other islands served as controls; their rats were left in place.

Adler wanted to know if the females would adjust their reproductive output in response to the changed density, and if it was male aggression that had limited their population. Adler suspected it was, because he'd seen the same effect in the temperate zone, in white-footed and deer mice.

After one manipulation, Adler noted that the "rate of density increase" went up: in response to the population decline — the subtraction of male adults — the survival of young rats increased. Next, Adler would return the adults to the exact spot at which they'd been caught and, with the sudden increase in males, enact the flip side of the experiment.

Adler planned to do the whole experiment again with females, to see if their aggression played any role in limiting populations. That's where Paul came in. After boosting the population with extra fruit, he'd help remove half the adult females from a subset of islands. He'd count the next generation, then bring the vacationing rats back home.

"Yeah, I surprised myself," Paul said, after he finished explaining the project. "I think I want to get more experience in the tropics. And then I'll start to think about grad school. I think I'm going on with biology."

I turned to Chrissy, who was sitting next to Paul. She looked tired, but she said that things had been going well lately. "You missed Derby Day," she said in a scolding tone, then gave me a shorthand account of the event.

"It was the first Saturday in May," she said. "We got up in time for a champagne brunch in the lounge. Everyone wore straw hats. And we had to drink bourbon and smoke cheap cigars." After brunch, residents regrouped at the end of Fairchild Trail for a game of water volleyball. The day's first race featured twenty dung beetles on a pile of monkey manure in the center of a dinner plate. "The first beetle to roll its ball of shit to the plate's rim won," Chrissy said.

At four o'clock, the action shifted to the helipad, out beyond the water-treatment building. The main event was the race of *Bufo marinus*, the giant toad. Twelve contestants lined up their animals, which they'd recently captured and trained. Did you wear gloves? I asked Chrissy, remembering that *Bufo*'s skin is toxic. "No," she said curtly. "You don't need to wear gloves." Even those who don't regularly pick up monkey dung bare-handed? No one wears them, she said. *Bufo* exudes toxins only when beat up, since poison is costly.

After all the residents had placed their bets, the animals sprang from the gate. Who won? "Some plant physiologist," Chrissy said, with a dismissive wave of her hand.

It was easy to get back into the rhythm of fieldwork with my regulars. When new people arrived, I got them to take me into the forest too. I attended weekly Bambis; I read my tropical biology. The heat and humidity bothered me during these first days back, and the forest seemed to be laced tight with spider webs. For this I'd left behind a temperate, lilac-scented spring? Such thoughts dogged me at first, but soon I was re-acclimated to life in the tropics.

One Tuesday toward the end of June I joined a half-dozen residents on the morning boat, headed for the weekly noon-time seminar at STRI's Panama City headquarters. I was hoping that someone had secured use of the BCI car, a dilapidated jeep parked in the Panama Canal Commission's dredging division lot, but we had no such luck.

Instead, we had the usual forty-minute death ride on a schoolbus crammed with kids, office workers, and canal laborers. The drivers seemed to delight in careening around corners on the narrow Gamboa Road, pushing the speedometer of the ancient conveyance till it felt like we were going eighty miles an hour.

The bus stopped a few blocks from STRI's collection of stucco buildings atop Ancon Hill, just inside the border of the Canal Zone. It was leafy and breezy here, with a great view out over the dirty, teeming city. The Smithsonian, which had moved peripatetically around Panama City since the 1950s, had settled here in the late '70s, on the site of the former Tivoli

Hotel. The hotel had been rushed to completion in 1906 for the arrival of Teddy Roosevelt — the first time a sitting American president had traveled abroad. He'd come to witness construction of the canal and to take a white-suited turn, for the cameras, at the controls of a ninety-five-ton Bucyrus steam shovel.

Today, the hilltop is home to two STRI administrative buildings, a multistory library, and a conference center named after Earl Silas Tupper, the plastic-container magnate, whose family donated $4 million to the cause. In addition to offices, the compound contains research labs, reference collections, a small bookstore, and a café. Scientists from all over the world stroll through this tropical campus, but the permanent staff can always recognize visitors from BCI: they're the ones with small red welts around their ankles. They're the ones who are scratching.

Noontime seminars, offered by visiting scientists and regulars, ranged widely through ecology and evolutionary biology. "Population Dynamics of a Social Spider" came quick on the heels of "Evolution of Parental Care in *Dendrobates.*"

The seminars interested me, but they also made a convenient excuse to get off the island, shop in the city, and work in the STRI library. With approximately 65,000 volumes, 1,000 journal subscriptions, and electronic links to databases in Washington, D.C., it was the best tropical biology library in all of Latin America.

That day I didn't pay strict attention to the lecture, on phenotypic plasticity in frogs. I'd been feeling queasy lately, and I kept falling asleep shortly after nine P.M. If I didn't improve soon, it looked as if Bret would have to train another triangulating partner.

With some shopping to do, I steeled myself after the frog lecture for the roiling city streets. I walked purposefully down the Avenida Central, shouldered my way into Machetazo, the

Panamanian equivalent of Kmart, and quietly asked the clerk for *una prueba de embarazo.*

I slept late the next day and, after checking to make sure my roommate was nowhere in sight, cleared a small space in the bathroom. Then I set up my rudimentary hormonal assay and watched in wide-eyed wonder as the single yellow stripe turned to blue.

The history of pregnant field biologists is not deep, although scores of female researchers have certainly climbed trees and counted coatis well into the later stages of pregnancy. The botanist Margaret Lowman even brought her infant into the field with her while measuring herbivory at twenty meters. When it was time to breastfeed her son, she came down.

But I was getting ahead of myself. In the first place, I was no biologist. And I wasn't sure I was ready to be a mother, although motherhood seemed like a good idea. What I had to do first, I thought, was clean up my urine experiment, making sure to hide the cardboard box in plenty of toilet paper.

I thought about calling Peter to share the startling news, but the matter didn't seem urgent. Instead, I walked to the end of Fairchild Trail. It was easy to put thoughts of pregnancy out of my mind. Things would be different soon enough, and I wanted things to be the same for a little while longer.

For nearly an hour I sat on the ground and watched the ships go by. I was grateful to be alone and grateful for the northerly breeze, ordinarily a phenomenon of the dry season, against my sweaty skin. I looked at the forest around me and tried to imagine, since I saw nothing but ants, the multitudinous creatures I knew it contained.

It was nearly noon now, and the usually raucous insects and birds had fallen silent. In a philosophical frame of mind, I imagined a rough parallel between the complexity of the forest around me and the complexity of the process within me. From there it was but a short leap to connect the ontogeny of this creature with the phylogeny of its ancestors. Human beings are said, by some, to be the epitome of vertebrate evolution; if so, then I had the potential to iterate that five-hundred-million-year-old vertebrate history. I could be a part of the process.

The notion seemed a bit grandiose. I was, after all, just another animal in a long line of animals. But there was something about being pregnant in a rain forest, a place seething with growth and fecundity, that made me smile. Finally, the forest and I had something in common.

Though I physically felt no different today than I had last month, I had a sudden realization, sitting at the canal's edge, that I wasn't alone — not in any sense. Everything in the world had come before this particular reassortment of genetic material; now, something would come after.

When I could stall no longer, I hiked back to the lab and called home.

11

Hubi, Toolmaker

FROM THE START I had noticed Hubi Herz around the dining hall and the lab. He was always darting in and out of the chemistry lab, popping into the lounge to speak to a friend, hurrying back out to his own lab. Hubi was twenty-nine, but he had the demeanor of someone older. He was focused on his work, serious. He walked with purpose, as if he had several appointments and his day was already backed up.

Being German, Hubi was part of the German clique, except that most of that group came and went and Hubi always stayed. He'd been on the island for several years and occupied a social rung far above the fray. He had a steady girlfriend on the island. He even ran his own lounge, referred to as the satellite bar, on the deck outside his lab. It consisted of a bench and an ashtray. Residents used it when they needed a quiet place to talk.

The flip side of Hubi's work ethic was his playful streak, his creativity. He was much in demand at Carnivalito, when residents broke into competing teams that designed costumes, sang tribal chants, and hurled antagonistic doggerel. Last Halloween he'd dressed up his lab as a space station: he'd programmed his computer to scroll intergalactic warning messages, fashioned a helmet from a bucket and some air-conditioning ducts, painted his rubber boots silver, and filled the room with dry ice. This coming October he would suspend a model of the swimming raft in the middle of his lab and scatter below it, on a bed of actual Gatun muck, the kind of stuff that routinely fell into the water: sunglasses, keys, and beer bottles, plus the data tables of one long-lost dissertation.

I wanted to go out into the forest with Hubi because he studied ants and, after spending time with Wilson, I had a bias toward people who study these little creatures. But it turned out that Hubi wasn't, strictly speaking, a myrmecologist. What Hubi studied was the interaction between ants and plants: as a member of the German plant physiology group, he was investigating how leaf-cutter ants, the genus I'd seen parading across the trail with parasols of leaf fragments, affect the trees that they harvest.

More than two hundred species of ants live off fungus they culture inside their nests. Among these creatures, about forty species make up the true leaf-culturing genera of *Atta* and *Acromyrmex*, while the other fungus-growing species thrive off cultures grown on dead insects and feces. On Barro Colorado, I learned, leaf cutters are among the most important herbivores. But no one knew exactly what effect they had on the island's plants. Hubi, in pursuit of his Ph.D., was estimating their herbivory. "I want to see what and how much the ants are taking," Hubi said, "and what happens to a tree while it's being harvested. I'm also testing whether the wounds that ants make on leaves impact the tree. Are they stimulating growth or are they stressing the tree?"

Hubi told me this as we knelt on the ground near Wheeler 3, our knees on thin squares of foam — chigger protection. (Hubi didn't use insect repellent; he said it would be counterproductive to working with undisturbed insects.) We were bent over a trail of *Atta colombica*, reddish ants with big heads. In Hubi's right hand was a Dustbuster, its snout modified to a point with a rubber hose.

He switched the machine on, and its whine, incongruous in

the rain forest setting, blocked out the background hum of stridulating insects and understory birds. After a minute or so, the vacuuming stopped. Hubi had what he wanted: exactly a hundred ants, holding in their glistening mandibles exactly a hundred pellets of debris. The pellets smelled loamy and looked like pieces of masticated corn.

Hubi emptied the Dustbuster into a plastic box and placed it in my hands. "For you I have a fancy Japanese tool," he said, slowly pulling from his rucksack the long black feather of a wild turkey.

My job was to rattle the box, knocking the debris pellets from the ants' grip, then repatriate the creatures to their trail using the feather. The leaf cutters were tenacious: some held on to their pellets even when suspended a meter above the ground, even when knocked against the box sides, even when jabbed with the feather's blunt end. Accidentally I sent a few into low orbit. "Could you try to put them close to the trail they came from?" Hubi asked. "I don't want them to get lost."

Hubi had an aversion to disturbing nature. If he moved a stick while collecting debris pellets, he made sure to replace it. He stepped over fallen palm fronds in his path, not onto them. Toward the streams of ants that flowed across the island, he was always a perfect gentleman.

While I bagged the debris pellets, which he would dry tonight and later weigh, Hubi did a sixty-second collection, this time without the Dustbuster, of ants carrying triangular leaf fragments toward the nest's entrance. He snatched the leaves with tweezers, but the ants, programmed over millennia to do only one thing, held on. Because Hubi was interested not in the ants but in the green material — he called it GM — that they carried, it was my job, once the ants and their leaves were running around the plastic box, to separate the two. Out came the feather.

Finished with this ant colony, Hubi packed up his tools — Dustbuster, feather, plastic box, two mechanical clickers for counting ants, tweezers, kneeling pad, meter stick, clipboard, notebook, and several Ziplocs. Like an anxious parent, he checked the nest area to see that nothing "weird" was happening, that the ants weren't removing a lot of dead workers or carrying leaves in the wrong direction. He counted all the incoming trails, he replaced a palm leaf that he'd moved while he was kneeling. Doctorish with his clipboard, he noted that the *Atta colombica* at Wheeler 3 seem to be enjoying fine health.

Henry Walter Bates, who cooked leaf-cutter ants into a spicy fish sauce, believed that attines carried leaf fragments as protection against the sun and used them to thatch the roofs of their nests and keep their brood dry. Leaf-cutter knowledge had progressed a great deal since then. Though not an obviously charismatic study subject, these ants and their sophisticated habits have long fascinated invertebrate zoologists, who have made of them a sort of intellectual pet, as writer Sue Hubbell has put it.

Ants, along with bees and wasps, belong to the order Hymenoptera, many of whose members are eusocial, or highly organized into castes. Each member of the colony is closely related to other members, so they tend to act as a collective, working for the good of one another.

Solidarity with one's kin turns out to be an exceptionally successful strategy. Hymenoptera make up two percent of the world's millions of insect species, but they represent eighty percent of its insect biomass. Ants dominate the rain forest, and they make themselves at home at nearly every latitude of the planet. "Wherever you go in the world, from rain forest to desert," wrote E. O. Wilson, "social insects occupy the center — the stable, resource-rich parts of the environment."

A phenomenon of the Neotropics, *Atta* colonies contain up to several million ants, all of them female (except during the brief period in which they reproduce), all of them the offspring of one fat, egg-laying queen. The rank and file are organized into castes that perform the chores that keep a colony alive and growing. Large workers clip fragments from leaves, then pass them off to medium-sized ants inside the nest. This caste chews the leaves into a pulp, but before they lay it on the fungus bed, they squirt the fungus with a drop of fecal liquid. It contains all twenty-one amino acids, essential for the fungus's survival. The fungus, of the Basidiomycetes group, digests the leaves' cellulose, a feat the ants cannot perform themselves. And then the colony eats the fungus.

The smallest caste of workers, called minims, nurse eggs or tend the fungus garden, cleaning and weeding it of any alien growth and keeping the culture pure. A caste of soldiers defends the nest; another removes exhausted leaf substrate — the pellets. Others play mortician, carrying the dead to the debris pile. Job descriptions among the three smaller castes are a bit fluid: as members age, they switch to a second group of tasks. Some minims ride atop leaf fragments to fend off parasitic flies.

In his elegiac *The Ants*, Wilson wrote that these creatures "represent the culmination of insect evolution." Indeed, leaf cutters not only invented agriculture some fifty million years before people did, but also figured out a way to exploit the abundance of energy in leaves by getting fungi to break down the terpenoids, alkaloids, and other sickening chemicals that tropical leaves use for protection against herbivores. Biologists have called the ant-fungus relationship one of the most remarkable examples of coevolution in nature.

With his rucksack settled between his broad shoulders, Hubi headed toward the next colony. Approximately fifty of them lay within a kilometer of the laboratory clearing. As the trail tipped downhill, Hubi's and my strides lengthened. Soon we were running, feet thudding on packed dirt. We turned off the trail at a scrap of flagging, dodging trees and fallen palms. We raced through a ravine, then slowed as we approached the next colony. Hubi didn't want to disturb the ants unnecessarily; he wanted a count of incoming leaf bearers that reflected normal activity, not the activity of a colony panicked by a galloping biologist.

Finding *Atta* nests didn't require great tracking skills. One simply followed the leaf-carrying stream. When two streams converged, the V pointed toward home. The nests were not cryptic: ants mounded up soil at their entrances and kept the ground above the colony vegetation-free. It wasn't known why they mowed, but Hubi suspected that it helped the ants maintain a low-humidity microclimate.

This colony covered perhaps thirty square meters. Some nests in the tropics sprawled over an area the size of a football field. Panting in the heat of the forest, we knelt by the nest's refuse trail. Hubi unpacked his counter and gave me one also, so we could compare numbers. He set his watch to beep in five minutes, wiped the sweat from his face, and laughed. "It's like a biathlon, isn't it?" he said. "You run and you run and then you have to sit very still."

Hearts pounding, we aimed our devices at the ants. Hubi set his watch. On the beep, the festival of counting commenced.

Hubi was sharp-jawed and lean, with green eyes that glimmered. He wore his blond hair in a brush cut, into which his girlfriend had shaved three skinny lines; they looked a lot like leaf-cutter trails. Hubi moved with the grace of a walking stick — the long-legged arthropod, family Phasmatodea, that resembles a twig. He never appeared to be walking quickly, but he could disappear through a tangle of green in an eye blink.

Fortunately, he was easily distracted. I'd catch up to Hubi while he was poking a stick in a web-covered hole. "Tarantula," he'd say. The animal would come charging out, curious to see who dared. Or a fantastically twisted liana would catch his eye. "Look at that," he'd say. "It's like art."

For two years Hubi had been preoccupied with measurement. He counted ants, he weighed debris, he calculated the growth rate of leaves. As others gestured in the air to amplify their words, Hubi drew graphs, with bell curves and poisson curves, straight lines and lovely forty-five-degree regressions that indicate one-to-one correlations. He spent his days collecting numbers and his evenings crunching them into significance.

Hubi filled me in on this one evening while we perched, rather formally, on stools in his lab. We were not drinking, and our feet were not surrounded by laundry. Things were not yet that casual between us. Hubi had certain boundaries that it would take a while for me to cross. His entire life was this island, and he tended to see me as an outsider, someone who came and went. Still, Hubi was happy to talk about what had brought him to Panama and what kept him on the island.

The youngest of three sons, Hubi had grown up in southern Germany's Allgäu region, surrounded by forests and dairy farms. Like any good naturalist, he spent his youth walking around in the woods, playing in the creeks. On weekends he hiked in the Austrian Alps, just thirty minutes away.

If Hubi thought of careers at all, he thought about forestry, or perhaps medicine. He was good with tools, he was inventive. But when his mandatory civil service came due, Hubi volunteered to band raptors and dig frog ponds for a local conservation group. After twenty-one months in the field, he made up his mind: he would study biology and, when he got his Ph.D., work in conservation.

He visited BCI for the first time in May 1993, as the field assistant to Rainer Wirth, a doctoral candidate studying leaf-cutter ants. "On my first day in Panama, I was invited into STRI's forty-meter canopy crane," Hubi told me, scratching his thatch of hair and squinting. It was his first experience of a tropical rain forest, and he vividly remembers his impression:

"It was super. Wheew! I saw termites working, I saw iguanas, all these different life forms. This is what I'd been reading about. I loved the lianas, climbing over these giant roots. Immediately I loved the climate. I'd felt it only in a greenhouse before; the steep angle of the sun — it was all so new to me.

"I rode *Las Cruces* out to Barro Colorado that afternoon. The noise of the cicadas had just started, and it was so loud. And the heat! I wasn't used to it, I was sweating all the time. I also partied a lot, that first week, and my first weekend I climbed the Big Tree."

The Big Tree was indeed spectacular, a *Ceiba pentandra* with a diameter of at least twenty feet. All island tourists were sent to see it.

"It was just like in Germany, where we'd climb every weekend," Hubi continued, "except here I was dangling on a rope, not walking up the side of a mountain. I spoke very little English then, and hardly any Spanish. I was never homesick."

When Hubi returned to BCI in 1996 to work on his Ph.D., he plunged immediately into the forest to locate his study subjects. He was using a map of the fifty-odd colonies made in 1993 by his predecessor, Rainer Wirth. (Like the bat group, the plant physiology group, which studied how plants cycled water and nutrients, had its own intellectual lineage on BCI.) But Wirth's map, he soon discovered, was useless. Almost half the colonies weren't there!

In some places Hubi saw a telltale mound of excavated soil with vegetation just beginning to regrow, but the ants were nowhere in sight. Aggravated, he'd move on to the next alleged site: again, no ants. By walking random transects through the woods, Hubi eventually found new colonies, but he wasn't pleased to spend his time this way. "I was mad that the lead guy's maps were so far off," he told me.

At this point Hubi interrupted his story to rifle through some papers on the counter. "Here's a picture of my home region," he said. It was a postcard from the Allgäu: a green dairy farm against a backdrop of blue mountains. Hubi looked wistful for a few moments, then continued his story.

His original plan, he said, was to start measuring the photosynthetic characteristics — the dependency on light, temperature, and carbon dioxide — of the ants' target plants, using his infrared gas analyzer. He'd plug these parameters into computer-simulation models that would in turn indicate how his trees reacted to the herbivory of leaf-cutter ants.

The gas analyzer, a $25,000 prototype he'd brought from Germany, was supposed to do it all. The machine contained a pump and a fan and a lot of onboard computers. It was covered with input and output shunts. Glass vials of soda lime, to remove CO_2, and silica, to regulate humidity, adorned its side. Slightly larger than a shoebox, the analyzer weighed ten kilograms. The cuvette, an electronics-filled tube that plugged into the central unit and gently held one leaf — still attached to its branch — weighed an additional kilo.

Hubi had high hopes for the machine and for the numbers he would soon be harvesting. He'd already attached a strap of webbing to its top using carabiners; he was going to haul the analyzer into the canopies of trees being harvested by ants and measure their immediate impact. He'd even lugged, all the way from Germany, a hang-glider seat, so that he could sit aloft for hours.

But right away there were problems. "The pump in the gas analyzer was broken," Hubi told me. "A regulator was broken, and the CO_2 dosing unit was broken. I was filled with enthusiasm and energy and I couldn't work."

He tried to repair the machine. Any day, he thought, spare parts from Germany would arrive. But they did not. In a fit of despair, Hubi went shopping.

He marched down Panama City's Avenida Central, the isthmian crossroads of all things polyester and plastic, and bought an inexpensive, low-tech device that he hoped would help him dose his target tree with CO_2 and then let him take the CO_2 away. He needed to establish a saturation curve — from low to high — that he'd plug into a mathematical model.

Back at BCI, Hubi unwrapped his purchase. He took a deep breath and slowly exhaled. His new biometric tool, a plastic Stegosaurus, took quite a bit of air. It stood about a meter high at the peak of its triangular brown plates. Attaching it to the crippled gas analyzer was a fairly simple matter, but out in the field Hubi realized he had another problem: he couldn't use the analyzer's valve to regulate both the flow of air from the dinosaur and its humidity. Soon he was forced to downgrade the Stegosaurus from an ingenious piece of lab equipment to a simple floating toy.

Hubi returned to his lab bench. It was no consolation to realize that even if the dinosaur had worked, he wouldn't have gotten far with it. The analyzer's light source was broken and the computer that ran the machinery was riddled with bugs. The analyzer, after all, was a prototype. Hubi had gotten it for free; now he was paying the price.

Fighting despair, Hubi switched his attention to another aspect of his thesis: were ants taking shade leaves or sun leaves? "Sun leaves are usually smaller in size than shade leaves," he explained. "They are tougher, have a lower water content and a higher rate of photosynthesis. They also have a higher nitrogen content, a different chlorophyll composition, and a higher specific leaf weight."

The ants' leaf choice mattered because a shade leaf doesn't suffer sun gladly; if ants were removing high, sun leaves, they were, possibly, stressing the leaves below them — newly exposed to the sun — and thus reducing the tree's overall ability to convert sunlight to energy. Or, in another possible scenario, the newly exposed leaves suffered little or no stress but compensated for the reduction in productivity.

To assess where the ants were harvesting fragments, Hubi needed to measure the distribution of chlorophyll within the canopy with a tool called a photometer, of which BCI owned exactly one. But the instrument was moldy and not nearly sensitive enough for Hubi's needs. New equipment was difficult to get — there were questions about who ought to pay for repair and whether purchases ought to be made in Panama or Germany or the States.

Months passed. Frustration abounded. Hubi walked the forest looking for field sights, preparing for the day when his gas analyzer would work. He began to monitor three ant colonies every three days for three months, noting which trees they harvested. It was a way to keep fairly close tabs on the entire harvest from beginning to end.

Hubi timed his visits to the colony's peak activity, making his own harvest, for one minute, of the ants' incoming leaves. Years before, Rainer Wirth had established through laborious counting that a day's entire harvest could be extrapolated from short-term measurements of peak colony activity.

"Didn't you mind spending so much time counting?" I asked.

"It was a little boring," Hubi said, "but I had to start somewhere." Counting, Hubi knew, was the starting point of science. It was the only way to know what was out there; it was the beginning of biological diversity itself. "When you can measure what you are speaking about and express it in numbers you know something about it," wrote the Victorian physicist Lord Kelvin in 1883. "But when you cannot measure it, when you cannot express it in numbers, your knowledge is of a meagre and unsatisfactory kind."

To extrapolate from the number of ants that entered a nest to the biomass of their harvest, Hubi needed to measure the average size and weight of their teetering loads. At day's end he brought his one-minute samples to the chem lab, where he sorted new leaves from old and flowers from stipules. He placed them on an area meter, a countertop device that calculates the square millimeters of any flat object placed on its circulating belt, and carefully recorded his numbers.

Others before Hubi had pondered the effect of leaf-cutter herbivory on plants. But to accurately assess the biotic mass that entered a nest was a Herculean labor. To get a statistically significant figure he'd need counts around the clock at considerably more than three colonies. "Still, the question interested me," Hubi said, "and I decided to pursue it." All he had to do, he knew, was come up with a more expedient method of counting. All he needed was a new tool.

As the weeks passed, Hubi watched ants enter their nests with leaf fragments — in addition to nongreen material — and exit with debris pellets. Just as humans locate their dumps on the outskirts of town, *A. colombica* dump their waste near the boundary of their colony, usually on its downhill side. (Other species of leaf cutters bury their wastes underground.) The road to the dump usually progressed along a promontory of some sort — a root or a liana — from which the ant unceremoniously released its burden. Picture a homeowner pitching an old rug into a landfill pit.

The ant trash accumulated spectacularly. One BCI leaf-cutter dump rose a meter off the forest floor — the myrmecological equivalent of Staten Island's Fresh Kills landfill, the tallest man-made landform on the eastern seaboard of the United States. In addition to dead ants, the piles also contained spiders and springtails and chiggers and beetles in adult, larval, and pupal forms.

Sometimes Hubi came upon ant dumps that looked wrecked. Any place that attracted so many arthropods was, naturally, going to attract arthropod eaters. Hubi suspected that coatis dug through the dumps. One day he had me fill large bags with dump material: he wanted to see exactly what the piles contained. "It's a whole system," he said, "but how does such a community get established? Are its members specialized? Is it deterministic? That is, can you predict the community in each one, and is there the same distribution from colony to colony?"

While pondering the dumps, Hubi noted that the stream of ants along the dump trail never seemed to vary, the way it did at nest entrances. He began to make five-minute counts at nests' back doors, hourly for twenty-four hours. His numbers were more or less constant. Hubi motioned for my notebook and drew me a quick graph. Plotted against time of day, the rate of debris dumping formed a perfectly straight line.

"Next," Hubi said, "I added up the integrals that composed the line, a figure that gave me the total number of refuse pellets dumped." In calculating the number of leaf-carrying ants that entered the nest, Hubi had already totaled the integrals that composed that bell-shaped curve. When it came to bringing in fresh leaves, it turned out, the ants' day, like many a human being's, started slowly, peaked about midday, and began to taper as the afternoon waned.

Hubi now had two numbers: one for incoming material and one for outgoing. He tested the numbers for statistical significance and came up with a P-value, or regression, of 0.0001. "That's significant," he said, with gleeful understatement.

The close correlation between what went out and what came in meant that Hubi didn't have to count individual ants all day and then measure leaf fragments at night; he didn't need to count leaf fragments at all. A five-minute count of refuse activity, at any time of the day, and a quick extrapolation gave him a fairly accurate picture of the number of fragments of incoming material.

Here at last was the tool that would do away with laborious measurements. Instead of measuring three nests in three days, he could measure fifteen in one day, or even twenty! It was like discovering that A squared plus B squared equals C squared. The formula would change the course of Hubi's dissertation.

Filled with a new sense of purpose, Hubi applied his novel method to the roughly fifty colonies that he was tracking. He minimized, as he put it, his efforts toward characterizing the ants' effects on trees — his original project. "I was just so excited about this new method of estimating herbivory; I was now studying ant behavior to get at the ecological impact on the forest, and it was too interesting to drop."

After thirteen months of counting, Hubi produced a number: leaf-cutter ants within one hundred hectares of Barro Colorado Island's laboratory clearing were eating 135 kilograms of biomass per hectare of forest per year. Each colony was harvesting 2.1 percent of the leaves in the roughly four-hectare area in which it foraged. Researchers had suspected that leaf cutters were the most important herbivore in a Neotropical forest; now, here was proof. They were eating even more biomass than the howler monkeys — whose diet was fifty to eighty percent leaves — in an equivalent area.

I was excited for Hubi, for coming up with an answer. But I wondered, also, what the answer meant. Did it lead science forward? Did it challenge any assumptions?

"Is Hubi's research novel?" I asked Bert, from the comfort of his floral-print chair. "Is it something you can fashion a Ph.D. thesis from?"

"Other researchers have asked about the effect of leaf cutters, because it's an important question," he said, sipping his whiskey. I had brought a water bottle with me; Bert was too polite to ask why I wasn't drinking these days. "But Hubi had the tenacity to stick with it and to come up with a new way of measuring. So yes, his work is significant." Precise measurement was near and dear to Bert, who was esteemed among biologists for his facility with mathematics.

Despite Bert's reassurances, it still seemed to me that Hubi had done an awful lot of work to get at one figure. But the figure was good, Bert had said. It was a puzzle piece that others could make use of. Like all basic research, it allowed other things to happen. And sometimes, I'd discover, it even put a new spin on ideas that had long been on the shelf.

Like other open-minded researchers — Bret, for example — Hubi allowed his imagination to be captured by all manner of interesting phenomena in the forest. His thesis may have started out narrowly focused, but the complexity of BCI continually tempted him to broaden his scope.

In 1996, at the beginning of his field season, Hubi had witnessed *Atta* carrying leaf fragments in two different directions over one trail. "I didn't know what was going on, and it was disturbing," he said. "The thinking is that ants carrying leaves tell ants coming out of the nest where to go. This bidirectional leaf-fragment transport didn't make any sense."

Was the colony moving? Or were workers from another colony raiding the harvest and brood of this one? Hubi couldn't call it colony movement until he'd actually witnessed the queen being moved from one nest to another. Once before he'd seen a colony move, in 1993, and had assumed then that it was the construction of a new dormitory, for the island's guardabosques, that had disturbed the ants. The movement was interesting to Hubi, but at the time he didn't think much more about it.

Now, years later, Hubi was watching colony movement once again: a trail of *Atta colombica* carrying neither leaf fragments nor debris pellets but, clamped inside their mandibles, ant larvae. He followed the trail backward, to an established-looking colony. Then he followed the trail forward, to a new-looking colony. The literature said that leaf-cutter colonies moved only rarely, and only under some sort of duress: construction disturbance, flooding, or perhaps fighting between competing colonies. Here, Hubi detected none of that.

Within a few weeks Hubi found another colony in transition, then another. Again there seemed to be no immediate cause. He became convinced that something significant was happening. "I began to count the ants going back and forth, carrying the colony brood," he said. "I was fascinated by it."

Slowly it dawned on Hubi that his predecessor hadn't done such a terrible job of mapmaking after all. Those original fifty ant colonies were exactly where he'd said they were — at the time. The movement of leaf-cutter colonies, then, was anything but rare. Which meant that whatever the effect of their herbivory on the trees around their colonies, that effect was mobile. For Hubi, it was eureka: 2.1 percent of the area colonized by leaf-cutter ants disappeared down the entrance of an *Atta* colony, over and over and over again.

For a place that had been under near-constant scrutiny for more than seventy-five years, Barro Colorado at times seemed to hold more mysteries than it had ever given up. Considering that entomologists founded the field station, and that leaf-cutter ants were the most common and prominent sign of terrestrial life around the lab clearing, I was surprised to learn that tropical biologists were unaware that these creatures, who were hardly secretive, were peripatetic.

Hubi now had a whole new area of inquiry for his thesis. In December 1996 he documented, for the first time ever, the movement of a colony's queen. For three nights he sat in the dark and pointed his video camera at a steady stream of workers bearing larvae and pupae. When it rained, the ants took cover along the trail, under an overhang or inside an earthen tunnel. One epic night, when the rain pattered on and on, the colony crossed two streams — once using a twig as a bridge, the other time a concrete block. These were, according to Hubi, moments of the highest drama.

To move a colony was no simple matter. Excavating a new nest was time-consuming and laborious: there was digging to be done, a new fungus garden and foraging trails to be established, new host trees to locate. Workers had to move larvae, pupae, and fungus dozens of meters. While the new nest was being prepared, the old nest continued to function. The whole process took up to six weeks.

The queen was among the last to make the move. She didn't just crawl out of her chamber and trundle down the trail to her new digs. She was a queen, after all: as the sole breeding female, she represented the future of the colony, the source of all its genetic information.

As Hubi filmed, the queen rode atop a single soldier. At rest, she was almost invisible, covered by a protective shroud of tiny ants. A fat, reddish animal about two centimeters in length, she once fell off her bearer — who was a tenth of her size. Workers rushed to cover her body. Once the queen was settled in her new colony, Hubi pulled out his tape measure. The distance between old and new nests was 130 meters. The videotape of this hegira was several kilometers long.

Colony movement was hardly the avenue of inquiry that Hubi's advisor expected him to pursue, but he did not want to relinquish it. "My grant was in plant physiology, from a group that studies stresses on plants and animals," he said. But the ants were beckoning him to explore a previously unknown animal behavior. Was there a way to unite the two pursuits? Others on BCI urged him to follow his instincts. "It's a golden opportunity," they said. "Take it."

Because Hubi's gas analyzer and his spectrometer were still on the fritz, and because the meager data he got using the Stegosaurus did not inspire confidence, Hubi bowed to the queen. He decided to include data on the movement of leafcutter colonies in his dissertation. Now, all he had to do was figure out what induced the ants to move.

12

The Disappearing Naturalist

T HE WEEKS PASSED, and I continued to count ants with
Hubi and track bats with Bret. I liked going out with
these researchers: they tended to get carried away with
the natural world. But sometimes I couldn't help but wonder,
yet again, if studies like theirs were an indulgence. It wasn't always easy to make the connection between a narrow-looking
project in the woods and a sweeping scientific finding on paper. The puzzle pieces, on the ground, too often seemed re-
dundant. Or too small to matter. Or specialized to the point
where they couldn't possibly interlock with others.

But then I met Doug Schemske, a botany professor from the
University of Washington, and the value of narrowly focused
research in tackling broad questions all of a sudden seemed
wonderfully clear.

Schemske arrived on the afternoon boat and came to dinner
late. He knew the routine: he'd been visiting BCI on and off
for twenty years, studying fish and birds and anything else that
captured his imagination. Looking weary, Schemske sat alone,
off to the side. I caught another glimpse of him in the morning
as he filled his battered rucksack with fruit. And then he disappeared into the forest.

It took me a few days to catch up with Schemske. Our meet-

ing was scheduled for eight o'clock one morning, near a particular plant that grew in the creek at Shannon 10. When I arrived, the professor was sitting in the creekbed with his eyes
focussed on *Costus allenii*, a member of the ginger family.

Standing about two meters high, the *Costus* had long, shiny
leaves that spiraled around its stem. Schemske studied three
species: *C. allenii* and *C. laevis*, pollinated only by euglossine
bees, and *C. pulverulentus*, pollinated only by the long-tailed
hermit hummingbird.

The *allenii*'s flower was red, about seven and a half centimeters long, and supported by a fifteen-centimeter green inflorescence, a rigid structure layered like a pineapple. "It's
green if the plant is pollinated by a bee," Schemske said. "It's
red if it's pollinated by a hummingbird." But it wasn't color
alone that kept the hummingbird from crossing the line: the
laevis and *allenii* had lobes on their flowers that physically excluded anything larger than a bee.

Schemske tipped the *allenii*'s flower to show me. "Bees arrive, tilt back to let out their long tongue, and dive down the
throat of the flower to suck the nectar. On their way in and out,
they collect pollen on their back." Each morning the *Costus*
produced one water-filled bloom, used it to lure pollinators,
and then jettisoned it by late afternoon. The whole process
started over again the next day. For something that was going
to be around for ten hours, max, the flower looked rather elaborate, but it was a relatively cheap investment, Schemske said,
considering what the plant got in return.

On this trip to Panama, Schemske was documenting who
pollinates whom, a steppingstone along his route to identifying the genetic basis for the characteristics that influence pollination. Was it the plant's flower color or shape? The place-
ment of its reproductive parts or the amount of nectar it
produced? Pinpointing which genes were responsible for these
traits, and in what order they evolved, would tell Schemske
about the number of generations it took to create a new species. Was it a thousand small genetic changes or was it a few

big ones? Through his narrow investigation, Schemske was answering one of the central questions of evolutionary biology: How do new species get started?

The minutes passed. Schemske sat on a rock and I sat on his canvas stool, our notebooks open on our laps. In the field, Schemske wore a faded and torn work shirt, khaki pants, and army boots. What remained of his hair was gray and longish. He kept his pencil extremely sharp. The creek splashed, sunlight dappled the rocky banks. It wasn't a bad place to spend the next eight hours.

At a few minutes after nine, the first bee came in. It hovered for a moment, circled, then dove into the flower. Nineteen seconds later, it left.

"What do you know about euglossine bees?" Schemske asked. "Because this is just a terrific species." He began to feed me, like the choicest of pastries, tidbits of euglossine lore. "Euglossines are a major pollinator of orchids. The males collect fragrances from various sources — decaying material, vines, orchids — and store them in sacs in their rear legs. Is it a precursor to a sex pheromone? We don't know. But the males seem to be selecting on fragrance, and the females are selecting on males."

Euglossines come in hundreds of different species, some the size of yellow jackets and some the size of gumballs. Their thorax and abdomen are iridescent with the blues, greens, and purples of an oil spill. Others sport orange and black hairs. The males live up to six months — positively ancient, for a bee.

With loaded scent sacs, male bees aggregate in sun-dappled forest gaps to buzz and flash their colors. The spectacle draws in female orchid bees, who seem to have no interest in a single scented male but will fly considerable distances to meet a crowd.

We got another bee in about ten minutes. Schemske noted its arrival and departure times. "These bees aren't coming in clumps," he said, excited. "It's not random and it's not constant. I wonder if they're all coming from the same hive. I think the bees protect the nectar for each other. Each one returns to the hive and lets the next bee know where to go. They keep out bees from other hives."

The plant, naturally, wasn't passive in the transaction that ensured its survival. The species *allenii* and *laevis* look remarkably alike; it is difficult to tell them apart, and they even have the same nectar production patterns, peaking between ten and eleven o'clock each morning. Schemske showed years ago that the two vegetatively different species evolved similar flowers, "cooperating, as it were, to provide sufficient flowers to command the attention of their shared pollinator."

Survival through mimicry is a theme nature plays again and again. It can occur in such sensory modalities as coloration, behavior, smell, or sound. For example, the viceroy butterfly, which is palatable to birds, has evolved the coloration of the unpalatable monarch. By riding the coattails of another species' warning coloration, it avoids predation.

Moths adopt the cryptic coloration of dead leaves or sticks. Nonvenomous frogs imitate dendrobatids, the frogs from which some indigenous peoples of the Amazon obtain poison for blowgun darts. Many nonvenomous snakes mimic the coloration of the poisonous coral snake. In the vegetative realm, seeds avoid seed-eating birds by disguising themselves as beetles or pebbles.

A fly buzzed the *Costus* but did not land. "The plants have an extrafloral nectary — a duct — near the top of the inflorescence," Schemske said. "They give this nectar to *Ectotoma ruidun* ants, which in turn keep out *Euxesta* flies, which want to lay eggs inside the flower."

Do the ants compete with the bees?

"No. The ants are getting amino acids from the nectar on the outside, and the bees are taking nectar only from inside the flower."

What had seemed like a simple bee-flower relationship was all of a sudden looking a lot more complex. This is what happens when you study one organism, intensively, for close to two decades. This simple flower, it turned out, was the hub of a teeming community. Schemske was happy to fill me in.

"The bees," he said, "carry mites, which jump off in the flower. They feed on the pollen, then jump back on another bee at the end of the day to get out. Hummingbirds also carry mites that feed on pollen; they carry them in their nostrils. I want to identify the mites and see if there's a differential — if species A is going to plant A, and species B to plant B." I foresaw another decade of research.

So far, Schemske's investigations into the adaptations and characters that contribute to reproductive isolation have revealed that *C. allenii* and *C. laevus*, though they share a pollinator, will not crossbreed; and that if the plants have different pollinators, as the *allenii* and the *pulverulentus* do, they can crossbreed, but they won't — at least in the wild — because nothing will bring them together.

Looking at the genome of F_1 flowers (the first generation of a parental cross), he discovered that one character — a leafy structure on the bracts that adds more photosynthate to the developing inflorescence — was controlled by just one gene. Usually it takes a suite of genes working in consort to have such a significant effect. "This is a pretty large-effect mutation," he said. "It was a real surprise."

In a comparison study, Schemske scrambled the genes of the offspring of two monkey-flower species, which grow in the temperate zone, and found that a single gene was responsible for fully half the nectar each species produced. "It was an incredible discovery," he said. "One species produced ten times more nectar than the other. And hummingbirds preferentially visited the flower with more — they detected the difference of one allele."

*

We watched the bees come and go. Once we saw two bees engage in aerial sparring that lasted ten seconds. "Phenomenal!" Schemske cried. "Just phenomenal!"

Up the creek, a video camera monitored a *Costus pulverulentus:* the tape would double Schemske's data set for the day. What was it like reviewing these plant videos? "It's a gas," Schemske said, beaming. "You see ants swarm in and fight with flies. Then a hummingbird comes along and they all scatter. It's a real big event."

Schemske was clearly a born naturalist: his knowledge was broad and he got excited about nature's tiniest dramas. His students and fellow faculty members held him in high regard for this, but the professor downplayed his expertise. "I don't know anything," he said. "I'm not even a botanist. I'm a fish person."

While the bees came and went, Schemske taught me the wolf whistle of the squirrel cuckoo and the songs of the chestnut-backed antbird and the black-footed trogon. He knew only twenty or thirty bird songs, he told me; the late Ted Parker, a tropical ornithologist, knew nearly four thousand. Twenty or thirty seemed just fine to me. I wondered how many of Schemske's students got to spend time outdoors with him. Every year fewer and fewer of them went into field biology upon graduation.

"Students burn out on nature," he said. "Maybe it's because they've been raised on twelve channels of nature programming. People used to sit around the dinner table here and go on about what they saw in the forest today, totally unprompted. You don't have that now."

Sometimes, though, residents did play at being old-fashioned naturalists. They hauled down the BCI logbook from a shelf in the lounge and recorded their observations of the day. It was a spotty record, but in one form or another the note-taking tradition had been upheld since the field station was founded.

The most recent volume began in the 1980s and served as a fairly good indicator of an eager visitor's state of mind.

Most entries involved animals, not plants. And the animals were usually doing something dramatic — dying, having sex, killing another animal. I feared it was the influence of those nature shows, with their inevitable emphasis on what the porn industry calls the money shot. Here, the money shot wasn't orgasm but a kill, or at least a carcass.

Then I checked out some earlier volumes and my theory was blown. It turned out that resident scientists had always been coolly obsessed with sex and death. I was drawn to the tireless observations of Russell Mittermeier, now the director of Conservation International. Mittermeier seemed to have spent most of his time on BCI, in the late seventies, walking the trails in search of carrion. In one day he came upon a dead agouti and a dead tamarin monkey. The next day's labors produced a dead armadillo, a two-toed sloth, and a peccary. "There seems to be a very high general mortality at this particular time of year," he noted. A contemporary of his wrote, "Saw two anteaters in reproductive act, meter 40 of Schneirla Path." Someone named S. A. Carey broke from the mold to report, "Spent delightful hour observing group of 'higher' primates feeding and competing to impress each other with their erudition and wisdom."

A few scientists tried their hand at natural-history illustration but most entries noted only the date, location, species, and activity witnessed. Lamenting this scientific minimalism, I came upon an exception, from John Pickering, a professor of entomology at the University of Georgia:

Pickering, 29 May 1995. On a sunny afternoon, Kathleen McEvoy and I swam from Gross Point to Fairchild Point. At about 4:30, when I was approximately 20 meters from the shore near the end of the point, about 50 meters from the beach, I felt as if someone was pushing on my left shoulder. Almost instantly,

I felt a hard blow to my left calf above the ankle. I turned around facing away from the shore to face my attacker. I had the horrible feeling that I was about to become part of a higher trophic level, realized that I was not only in deep water but in deep shit, and had probably pushed my luck too far. After a little tugging on my leg, which ultimately resulted in the loss of my Teva sandal, I reached down with my hands, intending to free my foot. At the same instan[t], my attacker released my foot and surfaced. Its jaw protruded from the water on my left side, revealing a fine set of teeth. The animal did not show its eyes, or at least I did not notice them. The mouth was completely out of the water, wide open. [The crocodile's] jaw extended above my eyes, probably about .3–.4 meters long and .25 meters wide at the base. I put my hand in its mouth to push it away. It went sideways and submerged, avoiding most of the force of my push. And that was the last I saw of the creature. A memorable experience that will last a lifetime. Elapsed time, possibly 5 seconds . . . I had 12 wounds on my left foot, 5 requiring stitches, 5 on my left hand, 1 requiring stitches, and scrapes on left side and neck . . . Just another adventure in the life of a field biologist. XO Pick

"Did you hear that?" Schemske asked quietly. "Motmot."

I'd missed the birdcall, my attention focused on what I thought might be a frog, sitting among some leaves downstream. We were into our third hour of observation now, and the more I learned about Schemske, of his devotion to this island and his broad interests — in fish, birds, bees, plants — the more he put me in mind of Frank Chapman, another well-rounded naturalist, who was pivotal to BCI's history.

The curator of birds at the American Museum of Natural History, Chapman had arrived on BCI in December 1925 and stayed for the four-month dry season, inaugurating a pattern that would last for the next twelve years. Thin-lipped and slightly avian looking, Chapman first chronicled his adventures in 1929's *My Tropical Air Castle* and then again in 1938's

Life in an Air Castle. These were popular books, and they were instrumental in raising awareness of the tropics as a place to conduct scientific research.

Like his naturalist forebears, Chapman had catholic interests: he wrote about pumas, ocelots, tapirs, collared and white-lipped peccaries, white-tailed deer, monkeys, sloths, tamanduas, tayras, and agoutis. He focused his more scholarly attentions on Wagler's oropendola, a crow-sized cousin to the oriole with a voice like falling water. Oropendolas build long, hanging nests in colonies; they select isolated trees that protect them against oophagous (egg-eating) monkeys but make it easy for birdwatchers to find them. They often share their nesting sites with hornets, which keep oophagous arboreal mammals, and botflies that could parasitize their babies, at bay.

For Chapman's "Bird Life of the American Tropics" diorama at the American Museum of Natural History, he drew inspiration and, literally, source material from Barro Colorado: he was a good shot, and he often took down specimens. During his time on the island, Chapman observed and listed more than 230 species of birds.

In the lab clearing, Chapman built a neat little house on stilts: two rooms, each twelve feet square, with a screened veranda upon which hung a hammock. As the years passed, Chapman added a back deck, where he sat with his spotting scopes, binoculars, and cameras, studying the oropendolas that lived in a tree behind the lab clearing. Under the house was a work area with a 360-degree view of the forest and the lake. This was his tropical air castle.

Chapman asked questions that researchers are still asking today, seventy years later. For example: What mammals live on BCI? Chapman suspected that the roster included most of the terrestrial species of tropical America, which meant that BCI was invaluable to anyone studying "the relation of an animal to its environment, physical and organic." Why were there so many animals? he wondered. What factors governed their numbers?

Unconstrained by grantmakers, timetables, or competitors, Chapman's mind roamed freely. Noting that brilliant and dull birds lived side by side, he asked, "Where are the forces of natural selection or rejection?" Protective coloration seemed meaningless in this context.

Turkey vultures intrigued him. Did they hunt by a sense of smell? He hid rotten fish inside a work shed, waited to see what would happen, and concluded, when the fish went untouched, that they did not. (A large and confused literature accumulated on this question, but it turned out that Chapman was wrong.)

To more readily study animal tracks, Chapman walked around the island with a child's rake and cleared trails of leaf litter. He cultivated non-native cassava and yams near the lab clearing before surrendering them to the omnivorous peccaries. A pioneer of wildlife photography, he lured carnivores to flashlight-rigged camera traps baited with meat (he photographed five different pumas in 1927) and kept a coatimundi named Juan as a pet.

Chapman was the Pollyanna of BCI. He found wonder in everything: the cockroaches that made their way into his house, a rogue coati that stole his bananas, flowering trees, rippling creeks. After a storm, "the sun, like a searchlight, sends down brilliant streaming rays and soon floods the earth with golden cheerfulness."

The island was, to Chapman, a scientific monastery: quiet, peaceful, and smooth-running. On BCI, he wrote, there are no "evidences of decadent human nature, in fact, none of the afflictions of modern life." Among such afflictions he counted automobiles, radios, jazz, movies, and "holdups." Apparently Chapman was unaware of the caretaker Donato Carillo's stream of female visitors, one of whom infected Carillo with syphilis.

His rhapsodizing aside, Chapman was not universally liked. The workers laughed at his Spanish, and Zetek confessed, in a 1925 letter to Thomas Barbour, that he didn't always make things easy for Chapman. But still, he wrote, "the old man is o.k. notwithstanding some of his queernesses. He wouldn't be Chapman if he was not queer."

Among Chapman's queernesses was a hatred of the sound of motors. Even slamming doors bothered him. He was no good with electricity, and one visiting scientist was certain he was going to blow himself up. When his "castle" slid down the hill in 1957, it was not replaced.

Chapman was proud to call himself a naturalist. He couldn't have known he was part of a dying breed. Science was changing: new technology was giving researchers better ways to look inside plants and animals — at cells, at molecules, at genes. Well-versed in mushrooms and birds, botany and geology, biologists like Chapman, who worked daily in the field, would soon be beaten down by competition for diminishing funds and by low morale, victims of the modern obsession with quantification and rigor.

In the January 1998 *American Naturalist*, Douglas J. Futuyma, an evolutionary biologist at the State University of New York, Stony Brook, wrote, "People [now] act like 'botanist' or 'ornithologist' is a badge of shame, and call themselves ecologists or evolutionary biologists, interested in their animals only as models for studying principles."

Up through the middle of the twentieth century, graduate students in biology could still earn their doctorates by describing species. Archie Carr, the great Floridian herpetologist and an expert on Caribbean natural history, for example, did the frogs of his home state. But in the 1970s, students began to receive training in the use of quantitative methods: in radiotelemetry and statistics, modeling and computers. Natural-history skills — quick: name that bird song! — were dismissed.

As scientists gained access to better computers, biometric tools, satellite imagery, DNA, and other molecular data, they spent even less time in the field. The change was reflected at universities, where courses in ichthyology, herpetology, mammalogy, ornithology, and the like dwindled in favor of courses in genetics and molecular biology. Field trips fell to budget cuts and liability concerns. Those who studied whole animals and those who studied molecules found their languages mutually incomprehensible.

With technology spitting out so much hard data, animal studies, which dominated the first five decades of Barro Colorado's history, began to appear soft, subjective. Many of the early vertebrate zoologists weren't well trained when they began to work. In the modern era, following a monkey or a sloth around, day after day, seemed more like outdoor adventure than forefront science.

Inarguably, animal studies lacked the rigor of chemistry or physics, disciplines defined by acute measurement and guided by immutable laws. Physicists can test their hypotheses using the principle of mass balance, for example. But organismal biologists have to make do with a tool set that is considerably more speculative: the fossil record, geographical distribution, convergent evolution, and comparative anatomy.

Running controlled experiments in nature — which is complex, messy, and changeable — is difficult. In the field, a data variance or an error of ten percent is considered insignificant. The challenge of recognizing factors that may bias results — the greatest responsibility of scientists — is compounded by the possibilities of subject perturbation or improper sampling.

Molecular biologists, concentrating in the sterility of their labs on just a few model organisms — yeast, viruses, fruit flies, mice — were soon "revealing an astonishing and deeply satisfying unity among life-forms in such matters as the genetic code, gene regulation, and biochemical pathways," wrote the evolutionary biologist and malacologist Geerat Vermeij in *Privileged Hands.* The breakthroughs the new technologies afforded, many of which had relevance to human health, attracted prestige and funding. (By the late 1990s, the National

Institutes of Health was funding thirty-five to forty percent of proposals for cellular and molecular work, while the NSF, after a period of invigoration, was down to funding just ten to fifteen percent of its ecology proposals.)

Animal behaviorists could hardly compete. Now and then someone described a new mammalian behavior, but the world did not take notice. Traditional biologists were, perhaps, ill equipped to handle quantitative science. By the early 1970s, wrote Vermeij, they "had jettisoned the data of natural history in favor of mathematically sophisticated abstractions."

But it wasn't high technology and higher math alone that changed the shape of field studies. In the mid-1970s, evolutionary biologists came under attack for explaining organisms' traits and morphology as the product of natural selection, or "optimal design." The two loudest critics, Harvard University's Stephen Jay Gould and Richard Lewontin, accused those who subscribed to the "adaptationist program" of glossing over evidence that didn't support their own ideas, of failing to distinguish a trait's current utility from reasons for its origin, and of relying upon plausibility alone as a criterion for accepting "speculative tales."

Gould and Lewontin made the case for genetic drift — that changes in some gene frequencies were random, as good as, or worse than, the next trait but here all the same. A trait could be the result of correlated responses to selection on a different character, or it could be an unmodified inheritance from the ancestral condition. The argument between drift and adaptation persists to this day, at a slightly lower pitch.

Twenty years ago, on BCI, the debate signaled a change. Vertebrate specialists, who did behavioral work, became scarce as emphasis in the universities shifted toward more quantitatively rigorous studies. Ecological studies, heavy on manipulation and experimentation, took off in the 1980s as the NSF created panels and enhanced funding in this area. By 2000, STRI employed only one vertebrate zoologist, an ornithologist who worked in Panama City and now spent most of his time indoors studying orchids. The institute's primary concern was now ecology and global change.

In order to survive in this new era, evolutionary biologists adapted: they learned the techniques of molecular biology. Some quit calling themselves naturalists. Within academia, biologists constructed their own pyramid of rigor. At the top were mathematical theory, because it most resembled physics, and experimental population genetics, which also relied on complicated theoretical models. Ecology and evolutionary studies malingered in the middle, with the stubborn animal behaviorists, taxonomists, and paleontologists relegated to the bottom of the heap.

The Stanford University biologist Robert M. Sapolsky called this reordering of priorities "physics envy," defined as "the disease among scientists where the behavioral biologists fear their discipline lacks the rigor of physiology, the physiologists wish for the techniques of the biochemists, the biochemists covet the clarity of the answers revealed by molecular biologists, all the way down until you get to the physicists, who confer only with God."

If scientists had the time and the encouragement, would they spend more time in the field? Schemske told me he didn't think so. Field biology is laborious and low-paying. Older biologists don't want to be away from their families, and in no other field is reward so far removed from effort. Like a seedling surviving the forest lottery, the biologist has her own series of challenges to surmount.

Before heading outside with a notebook, she has to devise a question, find out if the study has been done before, if the question has been answered or was found unanswerable. She writes a grant proposal, which spells out every methodological detail and every preliminary and expected result. Her proposal includes a phylogeny of her study organism, an explanation of how she spent her last grant, and a list of her recent publications. Then she waits half a year to get rejected. If she's among

the lucky ten percent to get funded, she completes her research, spends a year writing it up, then spends another getting
it published. By then six years might have passed.

No wonder so many biologists opt to stay indoors, where
systems are so much neater. "Now, people are content to study
a weed that's been around for just a hundred years and is pollinated by an introduced bee," Schemske said, still watching his
comparatively ancient euglossines in Shannon Creek. "They
sit in a lab and do population modeling on a computer. But
what can you learn from that?" Schemske was looking at organisms that coevolved over millions of years.

He went on, heatedly. "People genetically engineer crop
species for insect resistance, but they don't know the evolution
of the chemicals the plant started out with, how many mutations it took to get to that point. They don't want to know. And
they can't extrapolate their data to plants in the wild. They
don't know the natural history of the plant. It's just not fash-
ionable now to do the fieldwork."

Reed F. Noss, editor of *Conservation Biology*, addressed the
question of the disappearing naturalist in a 1996 editorial.
What is lost, he asked, when students are taught to model population viability and analyze remote sensing data but not to
"distinguish the song of the Bay-breasted Warbler from that of
the Cape May, the track of the mink from that of the marten,
the taste of the birch twig from that of the cherry?"

Noss bemoaned the loss of a personal connection to the
land, developed only after years of observing living creatures
in their natural environment. Without the emotional investment in a wild place, he asked, can a biologist exercise sound
judgment in making recommendations for conservation?
"Without years of bug-bitten trudging through hollows and
bogs, how can a biologist be expected to be able to separate biological truth from computer fabrication?" The environment,
in Noss's scenario, was the ultimate loser.

But do conservationists need to feel deep, soulful connections to the land in order to do their jobs effectively? I doubted

that. But naturalists are another matter. It has been said that good naturalists are born, not made. As a kid, E. O. Wilson mucked around in the Alabama swamps, dreaming of a government job that would let him drive a big green truck. Geerat Vermeij, blind since the age of four, combed the high-water line of Holland's beaches, collecting bagfuls of shells to sort at home. Charles O. Handley avidly trailed his father, a founder of the Virginia Ornithological Society, and his father's grad students through the woods. But with open space disappearing, where would the next generation of naturalists get their start? Would it be left to books and museum collections to inspire them?

Like indigenous tribes losing their shamans, who have intimate knowledge of the rain forest's pharmacopoeia, science has fewer and fewer people with the field skills to find and recognize creatures, especially in tropical forests. Taxonomists — who categorize, collect, and count newly discovered species, and who construct family trees that reveal how creatures live and relate to other creatures — are themselves in danger of extinction. The profession is barely encouraged in our universities. The crisis, Bert told me one evening, was hurting our understanding of what was going on out there. "You want to know how many different species of worm are in a tree, and what role each plays," he said. "If you can't tell them apart, you're in trouble."

About half the residents now on BCI were doing some sort of DNA work. It was trendy and it attracted funding. But some scientists — inevitably, those who'd earned their Ph.D.s without delving into genetic analysis — complained that the use of fancy equipment and new techniques short-circuited real thinking, that it kept scientists from asking broad questions about how the world works. In short, it kept them from walking around the forest and closely observing.

At the most basic level, there is still plenty to find out about the tropics. Some tropical forests have never hosted a collector. Millions of species are yet to be described. It doesn't take a multi-million-dollar lab to make a significant contribution to the field: open-minded biologists armed only with a notebook and binoculars are still turning up novel organisms and interactions.

And yet most of today's scientists choose to focus narrowly within their field, and they are realistic about why: generalists don't get jobs. It's sad, really, because it is the propensity to be impressed with all of nature, to be filled with wonder, that led so many scientists to biology in the first place.

I got a little wide-eyed myself in the presence of people like Schemske, who was both knowledgeable and enthusiastic about tropical forest. Like Handley before him, and E. O. Wilson and the great British evolutionist William D. Hamilton, Schemske is one of those rare biologists who have retained a sense of wonder at the complexity of the natural world while subscribing utterly to the scientific method.

Schemske moved easily between the field and the lab, ground-truthing his findings and refining his questions. Despite his fifty-odd years, he had no qualms about hauling loads of gear through the woods or sitting for hours in a creekbed. He actually liked teaching, which was rare in a research scientist.

Most impressive of all, he had mastered the modern analytical techniques — the better to put the puzzle pieces together — but he was not content to stay indoors and think. He had to go outside and look.

After ten days of watching hummingbirds and bees, Schemske returned to Seattle, where he was growing three hundred second-generation crosses of two *Costus* species, which he'd genotype after they flowered. The resulting map would tell him which unique traits were connected to which genes. In another year or so, he'd come back to BCI to look at a *Costus* species that made abundant seed without outcrossing — that is, without the introduction of pollen from another plant. The same species on the Panamanian mainland relied on pollinators. Schemske wondered: Was BCI, as an island, creating a new, self-pollinating species?

I called Schemske in Seattle to talk about his plans. "I'll work on these plants for as long as it takes," he told me. "It would be unbelievable to figure out their evolutionary history." I could hear him smile through the telephone wires. "And if I'm successful, I'll retire."

13

The Division of Cells

I WAS IN MY SECOND TRIMESTER now, and several mornings I woke before dawn feeling momentarily sick. It passed, and I dragged myself downhill to meet Craig, who'd replaced Paul on the spiny-rat traplines (Paul was provisioning the rats and living off-island now). Craig didn't eat in the morning, but I ate tiny amounts before going out, between islands, and on the way home. I drank water all day long — the mandate of pregnant women everywhere — and I peed on jungle islands, off the back of boats, and in the laborers' bathroom at the Panama Canal Commission's dredging division in Gamboa.

Through its ontogeny, the fetus had different names. Dr. Esquivel, the Panama City obstetrician to whom I'd gone for confirmation of my assay, had said that the baby, at two months, was the size of a rice grain. But I thought of the baby as a fish, and so that's what I called it. *Mi pescadito.* Sometimes, when I felt ambivalent about the creature growing inside me, I called it F_1.

My body was changing, but I tried to keep my schedule the same. I played Ultimate Frisbee a few more times, then gave it up. It was just too hot out there. Instead, I swam. Bret and I would strike out for the first buoy in Lab Cove, then head toward the second. I knew Gloria, the croc, was out here, but if Bret wasn't scared, then neither was I.

One day Bret showed me how to climb onto the roof of the dock and dive off. All of this was highly illegal. I got onto the roof easily enough, but I lost my nerve when I looked at the water, three and a half meters down. It took me fifteen minutes to slide off the roof and onto a pylon, my arms and legs were shaking so much. Was the fish trying to tell me something?

Back on the dock, I lectured myself: Fish do not talk. Was I losing my mind?

My husband, Peter, sent me the standard pregnancy text, *What to Expect When You're Expecting*, which stressed how "special" this time was for me. But I felt, here at the field station, just the opposite. Everything on the island was busy procreating, and no one made a fuss about it. Pregnant monkeys leapt from branch to branch without a visible hitch. Birds sat on their eggs, larvae crawled from galls on plant stems.

Because I was thirty-eight, the obstetrician had pronounced my pregnancy high-risk, but he told me not to worry. I probably wouldn't have anyway. Miscarriage, preeclampsia, and gestational diabetes never crossed my mind. I read the pregnancy book before turning in at night, which was a bit like reading Stephen King. The genre was horror — needles, incisions, stirrups — but somehow it wasn't my horror. I suppose I was in denial.

My emotions during the first trimester had swung wildly. At times I leaned my head on my desk and wept: What have I done? Thankfully, this weirdness soon evened out. Still, I felt no impulse to nest, as the book warned I might. My most consistent emotion was resentment, for my body was not my own.

It would have helped to have had an island confidante, but none of the female scientists had offspring, which ruled out commiseration. Phone calls back home were prohibitively expensive. And so I kept my own counsel, surrounded by rationalists, adherents of logic and falsification who, paradoxically, had far greater experience of gravid placental mammals than I did.

For whatever reason, I was unimpressed by the "miracle of life" growing within me. The zygote outgrew rice-grain measurements, its senses developed, my body changed. Everything about pregnancy was perfectly orderly. I was warned to expect absent-mindedness but spent more time pondering its adaptive advantage than wondering if I suffered from it.

In August, Dr. Esquivel, upon whose shelf sat *Que Se Puede Esperar Cuando Se Está Esperando*, remanded me, via e-mail, to the States for amniocentesis. In New York, a nurse practitioner looked at my medical records and asked me a few questions. "You're not digging around in the dirt, are you?"

"Yes, I am."

She made a note. "Well, you're not touching any animals, are you?"

"Yes, I am."

"What kind of animals?"

"I'm handling spiny rats and sometimes bats." I thought for a few seconds and added, "I also go out with a monkey researcher and collect fecal samples."

The nurse practitioner shook her head and scowled. "You wear gloves, don't you?"

"I will," I said soberly.

When I got back to Panama I told a few people that I was pregnant, and they became solicitous toward me. Chrissy wouldn't let me haul beer crates up and down between the lab and *Las Cruces* anymore. Rafael, with whom I worked on a migratory butterfly census, forbade me to carry full fuel cans. I was ripping into a bag of pumpkin seeds I'd bought in Panama City when Bret reminded me that seeds are often heavily defended against herbivores with toxic chemical compounds.

Among the alkaloids found in tropical leaves, seeds, roots,

shoots, flowers, and fruits, I learned, were cocaine, morphine, cannabinol, caffeine, and nicotine — substances that may cause cessation of lactation, abortion, or birth defects in mammals. Some plants contained saponins, which destroy the fatty component of the cell membrane. Others contained cyanogenic glycosides, which produce deadly hydrogen cyanide when combined with certain digestive enzymes. Members of the bean and pea family had toxic amino acids that interfere with normal protein synthesis. One of these toxic amino acids, concentrated in the seeds of some tropical plants and called L-DOPA, was a strong hallucinogen.

I went out on the *Urraca*, a fancy research vessel, but became so seasick in the Bay of Panama that I had to be let off the boat. I couldn't take scopolamine now. Nor could I use DEET to keep off the chiggers. Nor could I drink, enjoy the occasional cigarette, or stay up late. I was still visiting Bert on a fairly regular basis, but I'd switched from bourbon to water, or — if I felt deserving — a small glass of sherry. I wondered if the island's pregnant mammals altered their diet during pregnancy. Did their struggle to survive intensify?

Every few days or so, I'd rise in the dark and join Chrissy to follow monkeys and collect fecal samples from the two females I now recognized. I tried wearing gloves once, but they were hot and awkward. Besides, I felt stupid in them. The fetus, which had spoken to me on the pylon, let the matter ride.

Things were going well for Chrissy: the troop was acting in a predictable manner, and her mood was correspondingly bright. One night she dreamed that a female monkey came down from a tree and hugged her. As an expression of inner well-being, the dream was a classic. The next day, though, the dream appeared to be miraculously prescient.

Chrissy had spent the morning walking: she'd somehow lost the troop. Then she put her hands to her mouth and let out a "long call," a sharp, thin *Owwww*. The monkeys used this sound to signal their position over distances. Chrissy sat on a liana and repeated the cry. The long calling might affect the troop's behavior — it could wake them from naps or influence where they traveled — but she believed the benefits could outweigh the costs.

Before long, two male spider monkeys appeared in the treetops. One of the males, called A.J., climbed down onto the liana where Chrissy sat. She didn't know what to make of this. A.J. came closer. Frightened now, Chrissy moved sideways. So did A.J. He touched her lightly on the shoulder, then glided off.

Modern primatologists do not encourage physical contact with their study subjects. Jane Goodall and Birute Galdikas fed their monkeys to keep them nearby, but it made them more aggressive. Goodall's chimps began to travel in larger groups and sleep near her camp.

In the beginning Goodall didn't mind: she wanted to get as much data as possible from her animals before she had to leave them. She didn't know that her research would be long-term; the concept had yet to be born. Moreover, it was only by discovering something extraordinary that she could attract future funding.

"You must maintain a level of distance if you're going to study monkeys," Chrissy had told me. "And it's difficult with these animals because you have feelings about them. They're so funny to watch. They're energetic and interesting."

Chrissy was stunned, and also pleased, by A.J.'s touch. She had to admit, the moment had a certain poignance. Jane Goodall had been moved by a similar experience. Of the first time a chimp took her hand, Goodall wrote:

> At that moment there was no need of any scientific knowledge to understand his communication of reassurance. The soft pressure of his fingers spoke to me not through my intellect but through a more primitive emotional channel: the barrier of un-

told centuries which has grown up during the separate evolution of man and chimpanzee was, for those few seconds, broken down.

It was a reward far beyond my greatest hopes.

Chrissy didn't find the female spider monkeys after A.J.'s reassuring pat, but she came to dinner flushed and ebullient. When she told her tale to the table, the researchers oohed and aahed at what seemed to be a magical moment.

The hubbub subsided and — hormones being much on my mind lately — I felt compelled to ask, "Were you ovulating?"

I had some weird dreams of my own during this period. One early morning, in a hypnogogic state, I was in a race to pluck the fruit from two branches that met in a V. One branch was male, the other female; the fruits were alleles of genes. If I didn't reach the trunk of the tree by a certain deadline, rapidly approaching, I was going to throw up in bed. I was under tremendous pressure to choose carefully. But I couldn't. I woke up before I got to the V and made my dash for the bathroom.

Like residents who'd been here a while, I began to complain about the island's food. It wasn't ideal for pregnant people, especially if they didn't favor meat. The rice was white, the overcooked vegetables came from cans, and the cooks doused everything with margarine. Why couldn't we have brown rice? Because it's more attractive to vermin, I was told. Rats prefer brown rice to white, probably because it contains more nutrients. I worried about getting enough protein. I popped *acido folico* and, for the calcium, Tums.

Climbing a rope dangling from the branch of a *Sterculia* tree, forty feet up, I breathed in short gasps. I like to think I'd have made it to the top if I weren't pregnant, but the fetus was stealing my oxygen. There was a constant battle between my needs and F_1's — for my blood supply, for the nutrients I consumed and stored, for my air supply. The fetus was my own personal parasite, and I had few defenses against her. She would stay inside me until my placenta could no longer support her, and then she would be expelled. When my mood was sunny, I imagined that our relationship, from that point on, would be entirely mutualistic.

For the most part, I tried to ignore my physical condition on BCI and instead became attuned to the condition of others. I started to keep track of the residents' injuries.

In addition to Chrissy's broken arm and cracked head, early in her field period, she'd recently been chased, and stung, by an Africanized bee. Nothing untoward came of it.

A middle-aged myrmecologist from Switzerland broke his leg walking up to the fifty-hectare plot and spent his remaining three weeks on BCI hobbling about in a full leg cast. Trying to cut a piece of string with a machete, Sabine Stuntz sliced off the tip of her thumb.

And then there was my little injury. I held a *Leptodactylus pentadactylus*, a large frog known for its aggression — battling males grapple with their keratinous "thumb" claws, tearing into their opponent's flesh — and then rubbed my eye. The pain burned like a hot poker but subsided after a few hours.

Historically, the canal area was infested with creatures ranging from simply annoying to dangerous: boa constrictors, coral snakes, fer-de-lances, crocodiles, pumas, jaguars, mosquitoes, ticks, chiggers, spiders, and inch-long *Paraponera clavata*, ants that delivered a sting like a cattle prod. Early travelers faced parasites, strange viruses, and fungal and bacterial infections. Up to the time the Americans, with their mosquito eradication programs, began work on the canal, plagues of tuberculosis, smallpox, cholera, yellow fever, and malaria devastated towns and villages.

Today, most of this pestilence is irrelevant to BCI's residents. Mosquitoes are actually rare here: most standing water is inhabited by larvae-eating insects and fish. Still, the forest is home to plenty of stuff that can bite. There are urticating caterpillars, for example, whose hairs pierce the skin and cause itching, burning, and welts; there's the leptospirosis that Robert Stallard told me about; and Chagas disease, transmitted by assassin beetles, or kissing bugs, which congregated near the laundry room.

The most common island complaint was chigger bites. The lifestyle and habits of these tiny red arachnids dominated the small talk of newcomers. For those who'd been here awhile, this conversation had long ago worn thin. One's tolerance for the subject, then, was a de facto indicator of newness on the island.

Chiggers could barely be seen by the naked eye. Their natural host was reptiles, but they seemed just as interested in human flesh. One reference book said the larval form plagued us, another said it was the eight-legged adult. Insect repellent had some effect, but not enough. Taping the intersection of pants and socks helped, but it didn't prevent the creatures from biting torso and arms. If they made it past external sartorial barriers, chiggers tended to congregate near the elastic of underpants and bras, which caused some people to forgo these garments.

Where did the chiggers live, leaf litter or branches? No one on the island could say. Were they alive inside us, after they bit? Perhaps. How long could they survive on the couch in the lounge? For a place devoted to scientific research, a place enamored of arthropods, we seemed to have the least information about the creature that caused the most distress.

My injury list grew. A field assistant was bitten by a bat and had to get a series of rabies shots. An undergrad on the epiphyte team was nipped by a land crab while trying to photograph it. A woman examining how genes flow between *Tabebuia rosea* trees broke her collarbone when she miscalculated the recoil of the shotgun she was using to blast leaf samples out of the canopy. Every week Ultimate Frisbee's disabled list grew longer, with players suffering shin splints, pulled groin muscles, and dislocated fingers.

Craig came in from the field with wasp bites on his skull and around his eyes. "I got three stings, plus a coati turned over all my traps and stole my fruit," he said, pissed off. A week passed and Craig showed up, in an Earth First! T-shirt, with his eye swollen shut. A stick had slid into his eye socket while he was clipping rat toes. I reminded him of Earth First!'s slogan, "Mother Earth Bats Last."

Another few weeks went by; then Chrissy brought Craig to the hospital in Panama City, complaining of an irregular heartbeat after a night of drinking. The nurse at the hospital dismissed Craig's complaint and said, "What is it with you people from STRI? We had a guy come in with a sea-lion bite on his kneecap. We had a girl studying monkeys in here. She was screaming her head off. 'I'm gonna chase that monkey down! I'm gonna kill it and eat it for dinner!' "

"That was me," Chrissy snarled. The nurse slunk away.

By now I was working with Chrissy no more than four or five hours at a stretch. Round-ligament pain forced me to stop sometimes and rest. In the field with Craig, I had to turn my head when he clipped rat toes, especially toes that belonged to pregnant females.

If I didn't understand something that Bret said to me, he'd give me a quizzical look. "What?" I'd say. "Fish," he'd answer, nodding. It was his code word for the fetus, the fetus that was supposedly making me absent-minded. I felt like Ingrid Bergman in *Gaslight.*

I stopped fighting my fatigue: millions of years of evolution were at work, telling my body what to do. Now I came in from the field or the lab at four o'clock and collapsed on my bed. I read aloud to my little tadpole from *The Darwin Reader.* She flipped.

The tadpole turned, over the course of a few weeks, into a

dolphin (its ontological progress was erratic) and then rapidly into a chimp. She kicked and tumbled inside me. I began testing her, tapping on one side of my belly, then the other. She tapped back. Or was it the other way around? I couldn't tell if I was reacting to her or if she was reacting to me. As I well knew by now, designing a controlled study was no simple matter.

14

Wonderland

I LOST TRACK of Hubi for a few weeks. He flashed through the lounge now and then, and he came to dinner late, if at all. He seemed extremely busy. He was breaking in a new field assistant from Germany, and he had started collaborating with a Canadian researcher, billeted in Gamboa, on what induced ant colonies to move.

While Hubi was absent, I spent some time with Sabine Stuntz. Like Hubi, she wasn't technically a bug jock, but she spent her days with arthropods nonetheless. We met one afternoon on the balcony outside her lab so that I could help her build some traps.

With an Italian pop band playing on her boom box, Sabine sat barelegged on the concrete, surrounded by wires, string, Plexiglas, plastic buckets, grommets, pliers, and plastic sheeting. Her thumb was wrapped in a length of filthy tape, and she was twisting wire and tying knots awkwardly. The machete accident was making her job difficult, but it gave me a chance to step in and do some good.

Sabine, at twenty-six, was vivacious, dark-haired, and curvy, with a smattering of freckles across her snub nose. A German grad student, she was building traps, more than a hundred of them, to hang in trees and catch arthropods — ants, spiders, beetles, and other jointed-leg creatures. But she wasn't interested in just any arthropods: she wanted the ones that live in and among epiphytes, plants that wrap their roots around a branch but steal neither water nor nutrients from their host. Instead, they live off minerals diluted in rainwater. For her Ph.D. thesis, Sabine proposed that the canopy's great diversity of arthropods owes a great deal to epiphytes, which provide food for herbivores and increase the structural heterogeneity of the canopy — that is, they complicate the trees' architecture, creating more niches for more creatures. Epiphytes may also mitigate the harsh and arid conditions of the upper canopy. By trapping among these plants, Sabine was trying to find out the degree to which they influence relative abundance and species richness of arboreal arthropods.

For one year she would hang three kinds of traps — a flight interceptor, a yellow bucket, and a branch trap — in both epiphyte-laden tree crowns and "naked" control branches on twenty-eight *Annona glabra* trees spread over four mainland peninsulas. Others had assessed insect diversity and abundance in the canopy with insecticide fogging, but Sabine's study was long-term and covered more territory than those spot samples. To address seasonal fluctuations, the traps would be emptied at two-week intervals throughout the rainy and dry seasons.

Sabine had chosen *Annona*, also known as the swamp apple for its fleshy round fruits, because she already had baseline data on their epiphyte communities, and she could find individuals dominated by only one epiphyte species, which made clean comparisons possible. Moreover, the trees' roots were inundated through the rainy season and part of the dry, which limited the access of terrestrial arthropods. *Annona* grew at the water's edge and had a low canopy that, when Sabine first scouted her sites, she believed to be reasonably accessible from a boat.

Sabine was capable and strong. Back in Germany, she'd climbed mountains, snowboarded, and skated. In Panama, she wore short skirts, danced salsa, and attracted men the way flight-interceptor traps attracted *Polistes* wasps.

Visiting her field sites, some as far as an hour's boat ride away, was always something of an adventure. In the dry season we'd return from the field covered in mud. We'd pick ticks off our bodies for days. Once we were chased out of a tree by thousands of stingless bees, which crawled over our eyes and ears and up the sleeves of our T-shirts, seeking the salt of our sweat. Sabine had fallen from trees; once, she stepped out of the boat, barefoot, and nearly crushed a large red tarantula sitting in a tree crotch. But always, always, there was boat trouble.

It might have been bad luck, it might have been bad timing — too many users of one little boat. In any event, Sabine's hand on the throttle was not gentle. Sometimes she drove too quickly through low water studded with stumps, and sometimes she reversed the propeller too quickly when trying to free it of hydrilla. She'd been through one propeller and one transmission this field season.

One day we got back to BCI using our life vests as sails because the engine conked out in the middle of the shipping channel. We hit land at the end of Fairchild Trail, tied up the miserable boat, ate our lunch, and hiked back to the lab. "The breakdown sucks, but getting to hike is great," Sabine said. "Working on the lake all the time, I really miss the forest." We walked slowly, and Sabine taught me three new plants.

One morning at eight, when the rain was falling in thick gray sheets, Sabine looked at the sky and checked her watch. "It couldn't get much worse, right?" she said. We snapped our life vests over our rain jackets, which we zipped over garbage bags we'd pulled over our T-shirts. At the dock, we noted that someone had duct-taped to the steering panel of Sabine's newly repaired boat a saint card and some plastic flowers. "What a wonderful world," I said to myself. Then I got realistic.

"Do you have the radio?" I asked.

"*Ja*, right here," she said, patting her bulging rucksack.

We motored out of Laboratory Cove, rain stinging our faces. It was cold on the lake. Water streamed across the boat bench, into our pants on the left side and out on the right. Water poured down our calves and into our boots, where it sloshed between our toes.

It was a forty-minute ride down a winding finger cove to the first site. We kept our life vests on, for warmth. At the site, we poled through hydrilla and mud, then dragged the boat ten feet until we found something solid we could tie up to.

Moving awkwardly, we unloaded plastic tubs full of traps and tools. Then we began clambering up the *Annona*'s prop roots and trunk. Sabine had never imagined her fieldwork would look like this. During her pilot study she'd motored up to each *Annona*, reached overhead, and — twist, flick, pour, and snap — emptied and then refilled her traps. It had been the end of a normal rainy season, when water covered the trees' prop roots.

Now, because of the severity of the dry season, Sabine was playing monkey — stretching, balancing, hovering in midair as she worked with traps some ten or twelve feet above the lake's surface. We took turns: one of us would climb up and disassemble filled traps, the other would process the arthropods and refill the trap with one percent copper sulfate, a blue-green juice toxic to arthropods but, at this concentration, little else. It turned our fingernails green.

Was copper sulfate dangerous to a fetal human? We didn't know for sure, but we decided to play it safe. My days of mixing up killing juice were over.

Ordinarily I enjoyed this labor — painting fluon onto branch traps with a brush, tying string into guy wires. Sabine liked it too, though she said she was looking forward to the more intellectual aspects of her project, identifying the catch back home in her lab. For me, the trap hanging was art and architecture, with lethal intent. And then there was Sabine's running commentary, her enthusiasm for all things arthropodal.

"Oooh, look at this animal," she'd say, plucking a leaf hopper from a trap and admiring it from every angle.

Unfortunately, today's rainstorm was turning this excursion into an epic. The rain continued to lash us. Our mud-caked boots slipped over the wet branches. Perched on a high limb, waiting for Sabine to hand up more copper sulfate, I felt like an insect myself, at the mercy of some larger force. Off in the distance, lightning divinely flashed.

We finished the third tree in the group and headed for the next set, on Gigante Peninsula, but the boat immediately stalled. We removed the engine's cowling and wiggled all possible connections. I examined the starter, which spun when Sabine pressed the start button but didn't engage the flywheel. We fashioned a pull cord, which Sabine wouldn't let me pull, on account of my condition.

I'd outgrown my field pants by now, except for one pair that I fastened with a rubber band around the button. I felt fat. On another day when Sabine's engine had stalled, we had jumped out for a swim and I couldn't get back into the boat without jamming my stomach against the gunwale, and that hurt. Eventually we rigged a rope and she half pulled me inside. I found this coddling, especially from a younger woman, a little unsettling.

Now Sabine yanked as I stood by the throttle, ready to engage if anything should happen. Nothing did. With a sigh, Sabine unpacked the radio.

"*Me escucha BCI, habla Sabine, cambio.*" We could hear Cajár talking to someone else, somewhere, but he could not hear us. "*Me escucha BCI, habla Sabine. Cambio.*"

Nada, nothing: the battery was too low to send. We hauled the engine from the water and prepared to retreat under our own steam.

On the floor of the boat was one normal paddle and one meter-and-a-half-long branch, forked, to which someone had roughly tied a sheet of plastic, making a sort of water shovel. The lake here was shallow and clogged with hydrilla that slowed our progress and made us sweat underneath our garbage bags. We meandered past herons and other curious birds, under dripping branches, and around a dozen bends in the endless backwater alley. This stretch hadn't seemed so long when we motored in.

Every twenty strokes Sabine and I switched sides and paddles. We tried singing, but our breath came in rasps. We fantasized about our eventual rescue, out in the lake proper.

"Maybe the yacht crew will be eating lunch when they pick us up," Sabine said.

"Mmm. Some hot soup."

"A nice glass of wine."

"Maybe they'll have a hot shower."

"*Ja.* Thick towels. Dry clothing."

Eventually we fell silent again, and I asked myself the inevitable: What was I doing here?

After an hour we entered the lake and soberly noted the absence of ships. "It's so patchy," Sabine said, gazing off into the distance. She meant the ships.

"Patchy" was a favorite word among the biologists, usually used to describe the distribution of a plant or an animal — an irregular pattern of aggregation. Naturally, the word was incorporated into the everyday language of the island, like so much other scientific jargon. The beer machine was "heavily defended" — a resource to protect from rivals (or from removal by the administration). If you wanted vinegar on your fries, you had to go "foraging." Instead of saying a wild animal was fed, people insisted on saying it was "provisioned." When someone couldn't spot, say, a snake that someone else could, he'd say the snake was "cryptic," to excuse his lack of acuity.

I found this jargon amusing, and over the months I had slipped easily into using it. The way scientists talk had always been a big part of their appeal for me. Their language was precise, and they saw science in all that surrounded them.

Ten minutes ticked away. In the foggy distance, two ships passed. It was unlikely that they could see us. Optimistically we paddled toward a long finger of BCI, at least a mile away. Out here, the wind was even stronger, and pushing us backward. If we did reach BCI, we realized, we'd hit the end of Harvard Trail, a few miles from the lab, well after dark. Neither of us had brought a flashlight. The forest at night was inky black. We made for a buoy shaped like a Coke bottle and tied up.

Shivering, I wondered how long it would take to get hypothermic. We stared into the gloom, gloomily. I made a mental inventory of my backpack: compass, water, knife, map of BCI, notebook. In her backpack Sabine had a marvelous collection of string, glue, Plexiglas squares, and wire twist ties, but any shelter we could devise would fit, alas, only an insect. Sabine, a frugivore when in the field, had already eaten all five of her fruits. I was hoarding a Snickers bar, which I planned to eat at bedtime in the event of a nautical sleepover.

We were calculating the odds of anyone missing us when, out of the south, there arose the familiar buzz of a Boston Whaler. We sprang to our feet. A little brown boat appeared in the distance, making a beeline for BCI. "Wave!" Sabine and I screamed at each other. We swung our life jackets in the air. We stared at the brown boat, silent. It held its course and then, bing, it turned. "Yes!" we cried, slapping a high-five. It was Paul Trebe, returning from a distant rat island.

Jacketless and soaked, Paul tied our boat to his and ushered us into the Whaler. His fingers looked exceptionally white and prunish. I gave him my Snickers.

I sloshed up to the lab forty-five minutes later, feeling a little high from the misery of it all. Inside, a normal day was in progress. Bret was in his lab looking at pictures of porcupines. "Hey, don't forget it's Frisbee day," he said to me.

"In this rain?"

"It's a whole nother game in the rain," he said. "You're gonna love it. I promise you."

I shook my head and went uphill. I showered and started to pull on jeans, but then something overcame me, I don't know what. I had quit playing Frisbee because it was so hot, but today was cold. I dropped the jeans, dressed in shorts, and ran out to catch the Frisbee boat.

Through the dry season and the rainy season, Sabine's traps hung for two-week intervals. Then she detached each trap from its branch and poured the bug-killing liquid through a fine mesh screen into a bucket. The strained arthropods went into small bottles of ethanol, one for every trap, and flew back to Germany on Lufthansa, with Sabine, to be counted and identified.

During her first two trapping periods, at the onset of the rainy season, Sabine captured 36,875 arthropods belonging to twenty-five orders. By the time she tipped her ruined work boots into the garbage and yanked the last bit of string from her *Annona* trees, she'd have captured 267,389 arthropods from twenty-nine orders.

Sabine paid students at her university, and trained eight Panamanian students, to sort the arthropods to the level of order — Araneae, Hymenoptera, Coleoptera. Then she herself took over, first identifying the unknown creatures to the level of family or genus, then assigning them to morphospecies, or groups of arthropods with distinct physical characteristics. Hiring a taxonomist to get any more specific would have cost her more than buying BCI all new boats.

Sabine performed the most detailed inventory on spiders, because they are important invertebrate predators and highly responsive to the structural features of their habitat; on ants, because they are one of the most abundant insect taxa in tropical tree crowns and can heavily influence other insect orders; and on beetles, which are one of the most species-rich and ecologically diverse insect taxa in forest canopies.

Eventually, from her epiphyte harvests and from ferreting out the arthropods that live deep inside orchids and bromeliads, Sabine would discover that there were striking differences in species composition between different epiphyte species, a finding that suggests that epiphytes on trees promote higher beta-diversity, or differences between assemblages of species, for the entire area. In fact, she found almost no species overlap among the communities inside different types of epiphytes.

In one sweep, Sabine collected 1,700 ants belonging to thirty-four species. The number impressed her, though dipterans — mosquitoes and flies — would ultimately be the most abundant arthropods she captured in her trees. "Ants aren't the most species-rich creature in the canopy," she told me. We were in the boat again, on a dry day, straining arthropods from traps. "But they're very important because of their great abundance, because they fight off herbivores, prey on other insects, and consume nectar excreted by homopterans, or sap-sucking insects, that they farm."

Sabine never passed up an opportunity to impress upon me the larger significance of her work. She was keen on public education and dreamed of a job in the media, where she could share her knowledge of the rain forest. Now she added, "The total biomass of ants and termites together equals one-third the total animal biomass — that's mammals, reptiles, amphibians, and birds combined — in the Amazon." She waited for my reaction.

"That's a lot of ants," I said. I shifted my weight on the gunwale and flicked a drowned spider from my sieve. I knew where Sabine was headed with this: a lot of ants meant a lot of impact — on the breakdown of organic matter and the creation of soil, on the diets of ant-eating birds and mammals.

The implications were somewhat larger than Sabine's study suggested: she was focusing on arthropods associated with just three epiphyte species — *Vriesea sanguinolenta*, *Dimerandra emarginata*, and *Tillandsia fasciculata* — in one tree species. But nearly one-quarter of the plant species in a lowland tropical rain forest were epiphytes. The forests of Central and South America alone contained an estimated 15,500 epiphyte species.

*

Researchers have sampled arthropods throughout the tropics. To get a taste of the diversity in just one stratum of the rain forest, investigators fog the canopy with insecticide and rig sheets to collect the insect rain. Fogging by day and fogging by night yield entirely different crops. The Smithsonian's Terry Erwin, a beetle specialist who invented canopy fogging more than twenty years ago, has estimated that, in addition to the 370,000 known beetle species, there may be as many as 30 million more arthropod species awaiting discovery. Another expert's guess puts the figure at 80 million more. A single square mile of pristine tropical rain forest generally contains about 50,000 different arthropod species.

"We wouldn't even be talking about the rain forest if we didn't know how much was here," Sabine said to me. Still, I couldn't help but wonder, as I rehung a flight trap in a branch overhead: Now that we *do* have an idea how much is here, thanks to scientists like Terry Erwin, why must we keep studying the rain forest in such minute detail? It wasn't as if anyone were asking for more proof of the forest's intrinsic worth.

Just a few miles from these branch traps, visible from the back side of Barro Colorado Island, was a village called Las Pavas. There, people desperate for land were fighting the *paja blanca*, an invasive grass. They needed a place to cultivate their crops, they needed meat, they needed trees for firewood and houses. BCI had all those things. And yet BCI was held in reserve so that scientists could study whether damselfly larvae preferentially eat nonkin over kin. Some might ask if that money would be better spent figuring out a way to help the village, or the many villages, world over, just like it. Chasing data on arthropod diversity, in this context, could seem a rather elitist pursuit.

I considered the thousands of arthropods in glass vials at my feet. It was possible that some of them were new to science. Creatures with interesting family connections, pheromones, hormones, and habits. Pondering this, I realized that Sabine wasn't studying biodiversity to prove that the forest needed saving. She was studying biodiversity to act as its interpreter. That these creatures might have something to tell us was a profound and exciting notion. It reminded me that scientists study biodiversity, the living contents of the world, in order to know what we have and to parse the relationships between organisms. After all, it was relationships between organisms — the way they differed, competed, and adapted — that spurred evolution to reach the point that produced *us*.

Identified and scrutinized, Sabine's springtails and mites may turn out to have no immediate use for us. But their cousins, up or down the phylogenetic line, might. And if not, so what? Her creatures had been around for tens of millions of years — long enough to work their way into an intricate and self-sustaining ecological web. Their lineages had made it through glaciations and interglaciations. They'd beat out competitors for their particular niche, they'd lived through meteor strikes and earthquakes, deluges and drought. They were, in short, survivors, and they would probably be around long after the human race was gone.

A little respect was in order.

Eventually Hubi reappeared on BCI and the big rains temporarily ceased. I left him alone for a few days, and then I nabbed him.

"How are the ants?" I asked him.

"Meet me on the dock after lunch," he said. "Wear your bathing suit."

It was a bright, sunny day, and we set off across Lab Cove in a Boston Whaler that seemed just moments away from sinking. The engine kept stalling, and the plug that kept water from filling up the boat was not in the boat. Just beyond sight of the lab, we drifted to shore and hopped out into the mud.

Hubi wore a pink-striped shirt and flowered shorts, a pink life jacket, and a fluorescent orange cap. He reeked of sunscreen. "Today I'm measuring my artificial herbivory," he announced. We stopped beneath an *Anacardium excelsum*, which had large, waxy leaves. Known as the wild cashew tree, it has a straight trunk with deeply fissured bark. Howler and white-faced monkeys eat the tree's nuts, but humans have to roast them first to neutralize their cardol, a volatile oil.

"I damaged thirty-six leaves on this branch, and this other branch is the control," Hubi said, pulling down a branch decorated with flagging. The idea was to see how long the *Anacardium* suffered from its reduced leaf area and how quickly it recovered.

The leaves were numbered sequentially with waterproof pen. Hubi measured each leaf at its widest point and called out the number for me to record in his logbook. "Ten point one centimeters, eleven point three, eight point four, eight point two," he said, moving swiftly down the branch, from newer leaf to older. Later he'd calculate the area of each leaf individually — using a formula, tailored for each tree species, that required just the leaf width — and then add up the results.

"The question I'm asking here is how long it takes the tree to recover from herbivory, in either area or in the number of leaves. Because the leaves are the tree's productivity." That is, through photosynthesizing, leaves give the tree energy.

The numbers droned on and on, nearly a hundred for each tree. Hubi stopped to swipe an ant from his neck. "*Azteca*," he said, frowning. These small black ants live in the hollow pith of *Cecropia* stems and feed on its Müllerian bodies — nectar-secreting glands at the base of the petiole, where the leaf attaches to the stem. The ants' arsenal is both chemical and mechanical: after stinging their victim, the creatures rub a caustic homemade chemical into the wound. When crushed, the ants give off a bitter smell, redolent, some say, of gorgonzola.

Why does *Cecropia* tolerate and feed the ants? There are two schools of thought. The so-called protectionist hypothesis suggests that the extremely aggressive ants protect the plant, perhaps by keeping it free of vines and herbivores, and are paid for their efforts with nectar. The tree needs such mercenaries

because it doesn't invest in chemical defense; instead it allo-
cates its resources to rapid growth in light gaps.

The exploitationist hypothesis debunks this mutualism. It
argues that *Azteca* get a free meal while doing nothing in particular for their host. The ants, in this case, are mere parasites.

Whereas the *Cecropia-Azteca* question remains open, experiments on *Acacia* trees and their aggressive ants-in-residence,
Pseudomyrmex ferruginea, have shown that the trees do indeed
need the ants, to keep them free of herbivores and to clip back
plants that block their sun.

From the *Anacardium* we moved on to measure an *Inga gold-
manii* tree, farther along the shore. But first, some refresh-
ment. Hubi stripped off his shirt and jumped into the lake. I
followed suit; I loved swimming while pregnant. I felt much
more buoyant, and I liked the idea that the fetus and I were in
roughly the same environment.

Hubi paddled around, dove, and emerged with an interest-
ing-looking stick. It would join the natural-history collection
— shells, wasp nests, liana seedpods — that adorned his lab.
After swimming, we sat together on a sunbaked mud flat,
drying.

Hubi's fascination with the natural world was boundless, I
was learning. On the odd day that he forgot about leaf-cutter
ants, he and his girlfriend would take the canoe along BCI's
shore and beachcomb. He appreciated, volubly, the shapes of
various tree trunks, the perfect design of a chameleon. Like
Bret, he was always pursuing some peripheral project, just for
fun.

His most ambitious project to date was his grand tour. "It
was the end of the extreme El Niño dry season," Hubi began as
we motored toward the *Inga* tree. "More of BCI was exposed
than at any other time since the Chagres River was dammed, in
1904."

The most elaborate of his pre-trip preparations was the purchase of three pizzas at a Greek restaurant in town. He cut

them into pieces, slid them into Ziplocs, then set off to circumnavigate Barro Colorado Island on foot. "It was a challenge," he said. "I wanted to be the first to do it." I had been off-island then, but I might not have noticed Hubi's absence even if I'd been around. He liked to keep his private life private, and he was often missing in action.

Massive treefalls and extensive mud wallows extended the fifty-kilometer trip from an expected two-day tour to three and a half. Hubi slept in a hammock at night, under a tarp, and drank lakewater doused with bleach and a vitamin tablet for flavor. "It was awful," Hubi said, frowning at the memory.

We were at the *Inga* now, and Hubi measured while I wrote. Afterward, we motored back to the dock. I asked Hubi what he saw on the back side of the island during his trip.

"A crocodile twice the size of Gloria followed me along the shore for some time."

"Wow. What else?" We had collected our gear now and were walking up to the lab.

"I saw a tapir," he said, smiling. Tapirs had bristly manes, short, elephantine noses, and an acute sense of smell. Guardabosques knew of about seven on the island, but these hog-sized nocturnal ungulates were wary and not often seen by residents. (Alice, who had confounded Stallard's gauges, was long dead.)

Along his route Hubi had also taken note of otter scat, filled with fish bones and scales. "I remembered it," he told me, laughing, "because there was a scientist who spent a lot of time here looking at dung beetles and collecting various dung samples. But he never could find any otter scat."

Hubi's wide-ranging experience on Barro Colorado honed his innate skills as a naturalist. Loping between ant colonies, he would stop frequently to listen to birds, to pick up branches and rocks with a novel shape. Hurrying to a leaf-cutter colony near Bat Cove one day, he'd led me over a stream filled with small boulders. "Look," he said, bending over to make a tiny indentation in one of the rocks with his index finger.

"What is it?" I said, scoring the rock with a stick.

"Some kind of limestone. Don't mark it up," he said, shaking his head at me. He wanted to leave the area the way he found it. I dropped the stick and we hurried on to the ant colony.

Hubi had shown the rocks to Robert Stallard, the biogeochemist, who identified the mineral deposits as travertine. A few months earlier, Stallard and I had visited a travertine deposit farther downstream and watched the water cascade over rocks and into small pools. The rims of these pools, and their splash faces, were green and smooth and as drippy-looking as Gaudí architecture.

In a patient tone, Stallard explained the phenomenon. "The decay of leaves and the respiration of surrounding tree roots put carbon dioxide in the soil. It reacts with ground water to produce carbonic acid, which reacts with limestone to produce dissolved calcium. Working its way through the soil, the water ends up in the stream, where it degasses to the atmosphere. That process leaves the water supersaturated with respect to calcite, which then gets deposited out as travertine."

We knelt on the lip of a pool, following the dance of a small crab. "What makes it green?" I asked.

"Cyanobacteria," Stallard said. "They like this substrate. And when they photosynthesize, they also remove CO_2. So degassing and photosynthesizing are the two processes that get CO_2 out of the limestone."

I picked up a blue-green stone from a pool; it looked like a gem. The smoother stones were eroded and old; the rougher blue-green stones indicated a newer deposition.

"You'll notice that the shallower pools have more travertine at their edges," Stallard said. "Deeper pools don't have as much surface area to volume as shallower pools; less contact with the atmosphere means less outgassing. There's more outgassing on edges, which is why the travertine grows along the pools' rims and splash faces."

In the 1950s, a geologist named Woodring made a survey of Barro Colorado but failed to mention travertine in his notes.

Maybe, Stallard suggested, there was no travertine forming on BCI at that time. "But we know there was travertine earlier, because we see it under the tree roots. Either Woodring saw it and didn't mention it, or thousands of years ago, when indigenous people deforested this area, the soil was drier, it had less organic activity, and there was less CO_2 underground. We also know, from an aerial photograph, that this area was barely forested in 1927, though it's unclear when this intense deforestation happened. Between the late 1920s and now, we upped the carbon dioxide levels by letting the forest grow back and overtake the farms."

Would travertine deposit that quickly? I asked.

Stallard nodded. "It's a very fast rock."

It was useful to have a good geologist around, because BCI's rocks told interesting stories. They told us about the age of their setting, they told us about land-use patterns, and they told us where particular organisms lived. Once, for example, a visiting zoologist was puzzled to see what he thought was a cave-dwelling bird flitting around Lutz Ravine, and it disappeared before he could make a positive identification. Where were there caves on BCI? he asked Stallard. The biogeochemist consulted his files and pointed the birdwatcher in the direction of limestone deposits he'd earlier mapped. The zoologist came upon his bird — it lived in a tiny limestone cave just above Lutz Weir — and quickly made an important identification.

I found walking through the woods with Hubi — looking at lianas, lizards, and limestone — restorative, for I had become a little blasé about the forest in my months here. Hubi didn't know the name of everything he came across; he was a modern scientist, after all, a victim of the brutal niche partitioning that limits one's sphere of knowledge. But his energy and his enthusiasm for anything the forest offered did not wane.

I saw Hubi in Douglas J. Futuyma's description of the scientific naturalist: "the person who is inexhaustibly fascinated by biological diversity, and who does not view organisms merely as models, or vehicles for theory but, rather, as the raison d'être for biological investigation . . . the thing in itself, that excites our admiration and our desire for knowledge, under-standing, and preservation."

And so I admired Hubi. But after many months on Barro Colorado, I began to suspect that a fascination with the world's biological diversity may indeed be exhaustible. Even worse, I began to feel that the scientists of Barro Colorado played a large role in exhausting it.

It was partly information overload: scientists knew a great deal about nature in general, and Barro Colorado was particu-larly well studied. The gee-whiz feeling that I'd known as a child, alone in a forest, was quickly losing ground here to the minutely parsed and catalogued. Listening to a speaker explain the sex ratios of fig wasps or the relationship between root size and substrate structure, I wondered if scientists weren't exam-ining the life out of the island, peering so closely behind the forest's green curtain that it no longer contained any mystery. And I wanted mystery. Needed it, in fact. For without mystery, there was no wildness: the rain forest would be nothing but a big, wet diorama — a set in a theme park or zoo.

In *The End of Nature*, Bill McKibben suggested that nature is a diminishing entity. Physically there's less of it, and when we do go out into nature, we bring such accoutrements as cell phones and storm-proof tents that make "the wild" seem a whole lot less so.

Information does the same thing: geologists can roughly forecast earthquakes; entomologists tell us when locusts will descend; meteorologists can predict next week's weather — sometimes accurately! And so nature, measured and explained and predicted, loses some of its power over us.

And yet it was precisely that power, or mystery, that drove science and excited our desire for knowledge. As I spent more and more time among scientists, I came to realize that in the best of all worlds, a yearning for wonder and the drive for knowledge are not incompatible. In solving the mysteries of the natural world, scientists not only scratched at future mysteries but could also excite strong passions for preservation.

The biologist Karl von Frisch put it nicely. Pondering the ecology of honeybees, he wrote in 1971, "The more deeply one probes here[,] the greater his sense of wonder, and this may perhaps restore to some that reverence for the creative forces of Nature which has unfortunately been lost."

Hubi had just three months left on BCI. Finished with his regrowth-after-herbivory experiment in the field, he was now punching holes in plants in the growing house, covering them with plastic so their soil wouldn't lose moisture, then measuring the rate of their water loss per leaf area. In the field, damaged leaves had lost water much faster than undamaged leaves. But here in the lab, he was finding little difference between the control and the treated plants.

"Maybe it's because the plant is not water-stressed," Hubi said to me, perplexed. We were touring the growing house, just north of the labs, behind a scrim of low palms. The screened building held long tables of potted plants and was surrounded by a narrow moat, which kept out leaf-cutter ants.

Hubi explained that when a plant experiences water stress, it usually closes its stomata, through which the plant exhales carbon dioxide and loses water. Yet even when Hubi let his plants dry out, a major stress, he saw no big change in their rate of water loss. Had error crept in when he'd measured his plants' area?

"Someone told me that plant physiologists can rarely put their findings into ecology," Hubi said. "Their results are just in the lab." His tone was wry: field biologists liked to lord it over their laboratory counterparts because they actually observed organisms in their natural environment. (Later, after harvesting his experimental plants and properly analyzing his findings, Hubi would find a clear difference between plants with damaged leaves and control plants.)

I left Hubi in the growing house, puzzling over his lab results, and went for a walk down Fairchild Trail. I felt a need to
commune with the creature inside me. This was the first island
trail I'd walked alone, nearly ten years before, and I had a certain affection for its constant ups and downs. It was basically a
ridge trail, just 1,500 meters long, and it ended on a small
promontory overlooking Salt and Pepper Islands.

The trail was named for David Fairchild, who was on BCI
with Thomas Barbour and William Morton Wheeler when its
first buildings went up. Though he conducted no scientific investigations here, Fairchild made outsize efforts to keep the
lab from faltering in its early years. He gave his own money to
the station, and he traded on his family connections back in
Washington to win donations. Most important, he talked his
friend and benefactor Barbour Lathrop into doing the same.

It was Lathrop who first introduced Fairchild to the tropics. They'd met aboard a ship headed for Europe when Fairchild, an introverted midwesterner, was just twenty years old.
Lathrop was witty, handsome, and dynamic. He smoked Egyp-
tian cigarettes in a Turkish holder, knew everyone, and told
epic stories in which he usually figured as the hero.

The son of wealthy abolitionists, Lathrop had a keen interest in plants but no expertise in the field. (David Fairchild's
son, Graham, later wrote that he believed Lathrop to be a
"dollar a year U.S. spy.") Lathrop made Fairchild his protégé,
convincing him to study and collect fruits and vegetables of
economic importance and then to introduce them in America.
At Lathrop's expense, Fairchild traveled the world, eventually
creating for himself the job of Chief Plant Explorer at the U.S.
Department of Agriculture. But he wasn't often at his desk.

Where Lathrop was impetuous and bullying, Fairchild was
sober and polite. He had a boyish, enthusiastic laugh, and he'd
happily eat anything — noxious fruit, grubs, and even embryonic chickens. He loved the tropics and found them inspirational, a fail-safe prescription for intellectual malaise.

Over the years, Fairchild introduced many tropical and temperate fruits and vegetables to the nation's table. He played a major role in developing the American date industry and in planting Japanese flowering cherry trees in the Potomac River basin.

By the time he became involved with BCI, in the 1920s, Fairchild's bristling hair and push-broom mustache had gone white. In photographs taken on the island, he wears his shirt buttoned to the neck, and his eyes stare out from behind round, black-rimmed glasses. He looks stiff, but in chatty, self-deprecating letters to his friends he reveals himself to be afraid of large crabs, slightly clumsy, and something of a worrier. According to Thomas Barbour, Fairchild had the temperament of an angel.

I kept walking. Near Fairchild 3 I saw a woman moving swiftly through the trees to the north. I drew close enough to recognize Katrin, a German undergrad, who'd come to work with Hubi. She was slim and pretty, with her hair tied back in a ponytail, and she was carrying an enormous Hefty bag.

Months before, I had, at Hubi's request, filled several large Ziplocs with dump material — masticated fungus pellets, dead leaf-cutter soldiers, beetles, larvae, eggs, and whatever else these steamy mounds comprised. Hubi had noted the highly diverse arthropod fauna in the refuse material, but his main interest lay in the number of dead ants the mounds contained. The severe dry season had apparently caused a major die-back in workers, a phenomenon yet to be described.

Nothing came of this collection on my watch. Hubi was too busy counting harvested plant fragments and chronicling leaf-cutter movement to mess with this particular side project. But recently Katrin had contacted Hubi from Germany, looking for a diversity project related to ants. He suggested she work on the refuse piles, which was why she was lugging, that day, a Hefty bag bulging with sterilized leaf litter, leaf-cutter refuse, and sawdust imported from the mainland. With this material, she was constructing artificial dumps near ant colonies. What creatures would the dumps attract? Would dumps in different places attract different creatures, or would dumps made of different materials attract their own signature collection?

When she wasn't in the field with her fake dumps, Katrin sorted and identified arthropods she had collected from real dumps. The task was Sisyphean, and I had the feeling that she sometimes wondered what she'd gotten herself into. The dumps, Katrin discovered, contained mostly mites, followed by beetles, Collembola (a primitive insect order), Diptera, Pseudoscorpiones, Hymenoptera (or ants, bees, and wasps), and Blattodea — cockroaches. Among the beetles were several species new to science.

I waved to Katrin and left her behind in the woods. I was trying to keep up my regular pace, like the spider monkeys Chrissy suspected might be pregnant. At the end of Fairchild I longed to jump into the lake, but I was just twenty meters from the spot where John Pickering had been attacked. I took my usual seat, inhaled deeply, and looked around for my usual companions, the ants.

While Katrin pored over refuse piles, Hubi pursued other interests. After eighteen months of counting, tracking, and mapping *Atta* colonies, he'd found that slightly less than a third of fifty-eight colonies had moved. Which meant they relocated, on average, every five years.

Now the question that burned in Hubi's mind was: Why did they move? Setting aside the possibility of external disturbance and wars between colonies, Hubi and a Canadian grad student named Cameron Currie focused on pathogens. From the refuse piles of seven colonies in the process of moving, they collected pellets of dried-out fungus, let them grow for a few days in petri dishes, and then examined them under the microscope.

In a very high proportion of moving colonies, compared with nonmoving colonies, they detected a parasitic mold called *Escovopsis*. The pathogen was highly virulent, blooming like a white cloud atop the ants' garden of *Basidiomycetes* fungus. *Escovopsis,* Cameron found, could devastate a colony within a matter of days. The resident ants, if they didn't get out with their larvae, would starve to death. It was already known that when leaf cutters move, they pack up their fungus as well as their larvae and eggs. But workers cut only good fungus from their gardens, Hubi and Cameron discovered; they leave the infected stuff behind.

Next, Hubi and Cameron collected and examined fungus samples from newly founded colonies. All of the samples were "clean," except for one that came from a nest still under excavation. The biologists watched the project: the ants soon quit their labors and moved on, to start another, uninfected colony. "It was as if they realized they were still contaminated," Hubi said.

To test their hypothesis, that *Escovopsis* was driving ants to relocate, Hubi and Cameron infected several small leaf-cutter nests in Gamboa (such manipulation was forbidden on BCI). The experiment failed: no colonies moved during the brief observation period, although some ants died.

On BCI, Hubi and Cameron had seen ants living in nests that had only small amounts of the mold. "It seems they can tolerate a certain amount of infection," Hubi said. "We look at the refuse particles, and if less than one hundred percent are infected then the colony might be okay." The exact level of tolerable infection they'd attempt to determine in the future.

Though he hadn't planned on studying colony movement when he'd hauled his fancy gas analyzer and his hang-glider seat from Germany, Hubi had helped bring to the table another puzzle piece in the complex biology of ants. Cameron's *Escovopsis* finding was a nice piece of basic research, and it had an application as well. Leaf-cutter ants were a major pest in the tropics: their nests undermined buildings and their voracious leaf habit did more than a billion dollars' worth of damage to agriculture and forestry. Instead of controlling ants with

chemical poisons, which seeped into the water table, extermi-
nators might be able to use the *Escovopsis* parasite as a biological control.

After Hubi went back to Germany, Cameron Currie continued
studying *Escovopsis*. He wanted to know how leaf cutters sub-
sisting on a monoculture — a garden of just one fungus variety
— kept it free from disease. Monocultures, such as tree planta-
tions or single-crop farms, lack genetic diversity, which leaves
them vulnerable to pathogens and to changing environmental
conditions. Lower attine species, which grow their fungus gardens on dead leaves, insects, and feces, frequently change their
crop variety; but higher attines, the leaf cutters, stick with their
pure strain.

One day, looking through his microscope at a leaf-cutter
ant, Cameron made an important discovery. The whitish
growth on the ant's cuticle was not inert wax, as had been believed for a hundred years. The growth was, in fact, alive —
and teeming with *Streptomyces* bacterium, the source of fifty
percent of the antibiotics used in medicine. The bacteria had
little effect on most fungi but was deadly to *Escovopsis*. So important was the bacteria to the ants, Cameron found, that virgin leaf-cutter queens, like tourists heading for Mexico with
Immodium in their suitcases, carried it on their bodies when
they left to establish new nests.

The relationship between leaf cutters and fungi was hardly
unexplored territory. Biologists had been peering closely at
Atta nests for a century. That a graduate student could come
up with something so novel only showed the extent to which
the world is not known. The struggle between *Streptomyces*
bacteria, *Escovopsis* mold, and *Basidiomycetes* fungi — on the
battlefield of a leaf-cutter colony — hinted at untold intellectual riches yet to discover. For those who came up with the
right questions, who dared tread over ground believed to be
familiar, there was indeed plenty left to inspire wonder.

15

Saving the Whales

B Y NOW I FELT like an old hand on BCI. People were coming and going, I was assisting a half-dozen researchers, and every now and then a couple of tiny jigsaw puzzle pieces of research even fell into place.

There were jigsaw puzzle pieces in my head as well, several of which drifted together one Wednesday night. José Luis Andrade was giving a Bambi entitled "Regulation of Water Flux in Tropical Trees." With eight other researchers he had been working in Panama City's Parque Metropolitano on STRI's forty-meter crane, setting up the probes that Guillermo Goldstein had shown me in the lab.

In a breezy tone, José Luis first explained how the probes worked. Then he said, "To get the velocity of the water moving through the sapwood, we multiply by area and get liters per kilogram per unit of time. We found that the velocity is the lowest in the trunk and the fastest in the branches. The larger the diameter at breast height, the more sap the tree stores."

He picked up a piece of chalk. "Velocity times sapwood divided by leaf area equals transpiration," he said. Then he wrote on the chalkboard, $Gc = EP/Va$, where Gc is crown conductivity, EP is the amount of water, and Va is the difference between vapor pressure outside the leaf and vapor pressure inside the leaf. Then the questions began:

Are you subtracting gravitational potential when measuring water pressure in leaves thirty-five meters up?

Yes, I am.

Where are you measuring vapor pressure outside the leaf — next to it, or on the ground, or in Gamboa? (That was Bert.)

Right next to it.

The challenges continued, not as an attack but in a spirit of genuine interest: Did you include lianas in those figures? What about shade versus sun branches? José Luis had most of the answers, but many were "We don't know yet" or "We're going to look at that during our next study."

When Guillermo had shown me the hardware for the project, back in the lab, I had been thinking that studying how water moved through trees was a fairly arcane pursuit. I had felt like a heathen for asking, but I wanted to know if there was any practical application for such knowledge. Now, after several more months on BCI and a handful more Bambis, I was beginning to rise above my obsession with utility.

The central theme of Guillermo's project was understanding how a forest works. He and his team couldn't model a forest's response to environmental effects if they didn't know exactly where water traveled. With that information they'd have real numbers to devise real equations from which they could build their models. The numbers, downloaded from data loggers attached to trees in a Panama City park, were a tiny but quintessential piece of the puzzle. Guillermo was working on one of tropical biology's enduring questions: Are there organizing principles or laws that apply to tropical forests everywhere?

In light of global warming — which the STRI community fervently believes is happening and debates only the extent to which human activities are responsible — the basic studies conducted by Guillermo and the work being done on BCI's fifty-hectare plot were answering increasingly urgent questions: How does a forest respond to the changing climate? How does water move through a forest? Will forests be able to absorb elevated carbon dioxide through increased productivity? Will an altered climate cause a drop in biodiversity?

A molecular biologist appeared on BCI one night to talk about his work. By looking at fish genomes, he was tracing where populations came from, in evolutionary time, and — using their DNA as a fingerprint — where they went, in today's migrations. If populations were genetically distinct, they responded to different pressures and needed different sorts of protections.

When he finished his presentation, Nélida Gómez, a Panamanian Ph. D. who'd worked on and around BCI for many years, raised her hand and quietly asked: "With so little time to save these fish, how do you justify spending all this time and money studying their genetic differences?"

The room fell silent.

In the years since the word "biodiversity" had entered the lexicon, the years since I'd first set foot on BCI, little, it seemed, had changed for the good. Tropical rain forests were still disappearing at a rate of 144,000 square kilometers a year. For most of the 1990s, Brazil refused to enforce laws that protected its forests from illegal cutting. In the United States, 70,000 to 100,000 acres of wetlands were lost each year. Countless acres of grasslands had been converted to subdivisions and shopping malls; outside Phoenix, development stole an acre of desert every hour.

The United States failed to meet even the weakest goals for curbing greenhouse-gas emissions, and the Senate refused to ratify the 1997 U.N. Framework Convention on Climate treaty. Largely because of human activity, wild species were now going extinct at one hundred to one thousand times the natural, long-term "background" rate. One in eight plant species was threatened or endangered. Since plants form the basis of terrestrial ecosystems, the ripple effect of these extirpations on other organisms could be acute.

Environmental groups, which saw record-breaking contributions in the early 1990s, were now suffering from donor fatigue: Greenpeace had closed branch offices. Across the nation, Americans had overwhelmingly embraced sport utility vehicles, which emit two and a half times as much carbon dioxide into the atmosphere as do conventional vehicles. When he was the House majority leader, Newt Gingrich abolished the Office of Technology Assessment, which advised Congress on science and technology. His message, it seemed, was that scientists weren't worth listening to.

Were they? The question kept rearing its head. In the context of vanishing rain forests, why did knowing BCI's leaf-area index — the total area of leaves above each square meter of forest — matter? Or its fruit-to-frugivore ratio, or the amount of green material that arthropods consumed? How could one be a good environmentalist in this era of crisis and still believe that such narrowly focused studies in the tropics were time and money well spent?

I met a conservation planner in Florida who was bent on eradicating an exotic tree that threatened rare native species within the park he managed. But a scientist at the local university wanted to study the invasive tree for five years. By the time her study was done, the planner complained, her tree would have spread beyond any sort of control and killed off untold numbers of native species.

Was it better to take action? Or was it better to learn as much as possible about a threat before trying to vanquish it? The conservation planner and the university researcher were both scientists, but their agendas were radically different.

As I read the day's headlines on the Internet — Indonesian forests on fire, swordfish on the verge of extinction — the single-minded pursuits of the island's researchers again seemed to me esoteric. Residents traipsed around the lab in the usual Save the Whales T-shirts, but they didn't find it ironic that they weren't saving the whales; they were only studying them.

Nélida's challenge to the molecular biologist who studied fish threw me off stride that evening, because I had, after much introspection, come to terms with the value of basic research. (I wondered, briefly, if she'd challenged him because he was a *molecular* biologist.) Did Nélida have similar doubts about some of the projects pursued on Barro Colorado? I wondered. The fish biologist shrugged before he answered her question. "We make recommendations to the European Union, but no one listens," he said. "If we tell people not to fish, they tell us to go to hell." Then he added, in a plaintive tone, "I wish we biologists could do more than just learn things and have fun, but I think we have failed."

In mid-1999, Matthew Miller wrote in the *New York Times Magazine* that eighty percent of basic research is "idle, tenure-earning junk with little or no social value." His claim riled me. Even if it were true, one must ask how a scientist or granting agency can know what a study will uncover. How can anyone distinguish between the twenty percent that will lead to insight and the eighty percent that will go "nowhere"?

Charles Darwin spent eight years studying barnacles and fifty studying earthworms, simply because they intrigued him. His work may not have earned him an NSF grant, but his wide-ranging musings led him to his theory of natural selection. Basic research, then, is seed corn for the future. It's Cameron Currie scratching at ant cuticles to discover an important relationship between a bacterium and a parasite; it's Doug Schemske watching bees to learn how genes affect morphology.

Up until the 1970s, applied science, which is specific, and basic science, which is generalized, rarely overlapped. There was a prevailing notion that "smart" people did basic work. Graduates who went into economic entomology were ridiculed as "nozzle heads." But ecological studies changed much of that. Today, those who practice basic science — studying an organism's anatomy, say, or its behavior — and those who practice applied science collaborate splendidly, in pursuit of conservation.

José Luis's data on the movement of water through the rain forest may have utility to the Panama Canal Commission or to farmers in the Canal Zone, but a complete understanding of this ecosystem echoes around the world. Now covering six percent of the earth's land mass, rain forests have a significant effect on global systems — on rainfall, on climate, on migratory animals and the creatures associated with them. Including humans.

The world's poor currently number about 1.1 billion people, the majority of whom live in the tropics. If the forests that support them — supply their water, their timber, their food — unravel, as many believe they have already begun to do, the migration of these populations would have untold impact on developed nations.

Because rain forests affect everyone, is this, then, a reason to study them? If preservation is the central goal, wouldn't it make more sense, as Bret had said, to simply buy the largest plots of the most vulnerable forests and give disenfranchised locals a reason to protect them?

No, I was learning: setting aside land is not enough. In North America, it was becoming apparent that for conservation to succeed, we need not only to preserve land but also to know how to manage or even restore it. When developers are allowed to construct wetlands to mitigate the negative impact of their shopping malls, they need to know, for instance, how many acres the resident amphibians require to breed successfully. They need to understand the intricate mutualisms that make breeding ponds hospitable.

Development schemes must also consider the life histories of the organisms they may impinge upon. During the rainy season in the Brazilian Amazon, fish swim through the forest's bottomland, feeding on fruits and seeds that drop from trees. In turn, the fish provide a major source of animal protein for the forest's human inhabitants. "Had the forests been cleared before we learned about the fishes," wrote Forsyth and Miyata in *Tropical Nature*, "this valuable resource would have been lost without a clue as to why."

In 1998, a team of 119 researchers questioned the value of habitat-conservation plans in protecting endangered species. "Many of these plans lack the data to do anything even remotely scientific," said the zoologist Peter Kareiva of the University of Washington, who oversaw the study. "If you don't know the basic biology, it's sort of a delusion to think you're doing anything to help these species."

And so we study biodiversity because biodiversity has things to tell us about how ecosystems are assembled. In the last fifty years, scientists have recognized that organisms — tropical or temperate — are exquisitely, and often quite unexpectedly, interlinked. Human impacts on the environment can therefore reach much farther than is at first apparent.

In the late 1990s, researchers at the University of California, Santa Cruz, noted that sea otter populations in the Pacific Northwest had declined ninety percent since the 1970s. Through radiotracking, the scientists were able to rule out reproductive failure as a cause. Lacking corporeal evidence — dead bodies — they dismissed starvation, pollution, or infectious disease.

For decades the researchers had seen orca whales swimming peacefully around otters, but in recent years they had seen orcas killing otters. Why had the relationship soured? In the journal *Science*, the researchers proposed that orcas were eating otters because their usual prey, seals and sea lions, was scarce. Why? Because *their* usual food, perch and herring, had lately declined — owing either to overfishing or to competition from predatory pollack, which proliferated when whalers reduced the population of whales, which live off microscopic animals they filter from the water. As these organisms multiplied, so did the pollack that gorged on them.

To summarize: Whalers took out whales, plankton proliferated, pollack boomed, perch and herring went bust, seals and sea lions followed, and so orcas switched their diet to otter. As the disappearing otter has a back story, so too does it have a front story, a cascade of effects rippling into the future.

Without sea otters plucking sea urchins from kelp forests — great coastal stretches of algae — the urchin population booms and quickly "deforests" the kelp ecosystem. With less algae around, limpets, sea mussels, and barnacles decline, as will the resident rockfish, sculpins, and greenlings. Bald eagles, partially dependent on these fish, could also become locally extinct.

When the sea otter researchers began trolling two thousand miles of ocean thirty years ago, they weren't trying to save the national bird or even kelp beds. They studied sea otters for their own sake. Their research was basic; it had no application to mariculture, to medicine, to conservation. And yet it solved a biological mystery of broad proportions, and it may even help to save a species.

By studying species for their own sake — a time-honored scientific tradition — researchers have discovered that some creatures act as sentinels, telling us about the health of the ecosystems in which they live. For example, herpetologists have found that frog and toad populations from the Americas to Australia have been drastically reduced within the last several years, even within protected areas. With their permeable skin, amphibians are supremely sensitive to disturbances in the air and water. They are one of the first animals to disappear when habitat is threatened.

Biodiversity studies are also a form of exploration. Much is made of the tens of millions of insects that remain undescribed (taxonomists tackle about 7,000 newly discovered insects a year), but species even larger elude science's grasp. Since 1977, more than 150 new mammals have been identified. In the last decade, zoologists have discovered three new species of deer, called muntjacs, in Southeast Asia, a region long kept off-limits to scientists by war and politics. Even creatures common to our North American backyards aren't entirely understood: some male beavers, it was recently discovered, have a uterus.

Biodiversity also tells us about carrying capacity. Researchers periodically conduct animal censuses on BCI and in other nearby nature reserves. They want to know what is out there, to identify the causes and magnitudes of population fluctuations and — here comes the utility — to gain information about "minimum viable populations"; that is, how much a community can shrink before inbreeding, disease, or stochastic events, such as hurricane, flood, or drought, can do it in.

How exactly are these data useful? In a world where natural vegetation is increasingly confined to reserves or forest fragments, knowing how much of the original biota — plants, animals, and other living organisms — can be preserved in a park of limited size is valuable information indeed. Islands make excellent systems for studying minimum and maximum critical populations, the effects of invasive species, and the dispersal of plants and animals.

The late Robert MacArthur theorized that the great diversity of tropical communities reflected the great stability of tropical environments. If this is true, then the fluctuations of plant and animal populations, and their causes, were of significant interest. The data could act as the operating instructions for restoring or managing tropical forests.

While many scientists on BCI connected their basic research with conservation — in the "If we don't know how it works, how can we protect it?" vein — others had no illusions. They considered conservation a matter of politics, not biology. Basic research, useful or not, was what you did to earn a degree.

À Ph.D. was an enfranchisement, in this scheme: it gave a scientist a place at the table. Not all the scientists in training on BCI wanted to continue as researchers, in academia or outside of it. Some were interested in influencing government or social policy.

"With a degree, people will listen to me," Nigel Asquith, a British doctoral candidate, told me, just as Bret had. "They'll know that I know what I'm talking about because I've been in the field, running experiments, and I know how these systems work."

But would residents of the tropics welcome Asquith's advice? The reality was that few developing countries had the luxury of seriously addressing species or habitat conservation. They lacked the funding, and they had more immediate concerns. For all the talk of conservation around BCI, temperate-zone ecology was far more conservation-oriented than tropical ecology.

Still, people like Nigel and Bret, with their hard-won Ph.D.s, would soon be in a position to share their knowledge and opinions. Even if performing basic research in the tropics did nothing to preserve the region's biodiversity, at least it had enriched these scholars' minds. And presumably, we'd all be the better for it.

A few days after José Luis's Bambi on regulation of water flux, I walked down Standley Trail, in the island's northwest quadrant. Paul C. Standley had been the assistant curator of plants at the National Herbarium, and he'd written the first flora of BCI in 1927. It was my first time down this trail — it started several miles from the lab clearing — and I'd been saving it as a reward for staying on the island so long.

My reward became sweeter almost instantly. Not too far down the trail I came upon two collared peccaries, which look like wild pigs but belong to their own family, the Tayassuidae. Their common name comes from the yoke of white hair around their shoulders. I'd often seen loose soil on other trails, churned by cloven hooves, but I'd never run into a peccary herd, which could contain up to thirty animals.

Not long after I saw the peccaries, a low-slung, bushy-tailed animal with a face like a bear's streaked across the trail. I was having a banner day. The creature was a member of the weasel family — a tayra, or *Eira barbara*. Two-thirds of a meter long, not counting its tail, the tayra fed omnivorously, day and night, and lived in hollow tree trunks.

One could become jaded about the forest if one walked it daily, but this little stroll reminded me how lucky I was to be living in a nearly intact rain forest. How many people, after all, have ever seen a tayra? Or a living blue morpho? Or a full-grown mahogany tree, like the one that grew a few miles away in Gamboa? Nicholas Smythe, a STRI researcher, had seen the tree filled with nearly three hundred red-lored parrots. Actual wild parrots!

Writing in a 1992 STRI newsletter, Smythe reported that a grison, another member of the weasel family, which had never been studied in the wild, was killing chickens in Gamboa. A bee researcher had seen a jaguar cub playing in the middle of that town's Pipeline Road. "Jaguars are one of the most endangered animals in the world, to see one at all is a lifetime experience; yet here we have good evidence that jaguars are breeding within an hour's drive of Panama City," wrote Smythe.

When professional conservationists speak publicly about the value of an intact biota, they present a litany of the rain forest's biodynamic potential: as a source of food, fibers, energy, genetic information, and pharmaceuticals. Forty percent of our medicines, for example, contain at least one compound obtained from rain forest plants. Physicians prescribe anticoagulants based on leeches and on snake venom. We get leukemia treatments from rosy periwinkles, and treatments for Hodgkin's disease, heart ailments, and arthritis derive from other tropical plants.

Healthy ecosystems also provide the services that allow humans and other creatures to survive. They clean and recirculate air and water, mitigate droughts, floods, and temperature extremes, decompose wastes, control erosion, create new soil, and pollinate crops.

But utility — what intact ecosystems are worth to human beings — isn't everything, and now and then someone takes the high road to defend rain forest conservation. In his introduction to *Tropical Forest Ecology: A View from Barro Colorado Island*, Bert Leigh wrote, "The tropical forest will survive only if we recognize its value over and above its usefulness to ourselves, just as scientific research will continue to bear useful technological fruit only if some research remains motivated by an interest in nature unrelated to its usefulness to ourselves." Leigh's is, essentially, an ethical argument: Save the rain forest and all it contains because it is there. Because it is wondrous and interesting. Because it has a right to exist.

The more I contemplated what the loss of such places felt like to me, the more I began to think that some of these "higher" reasons for conservation might be just as utilitarian as the food and gas arguments. Wasn't the wonder of seeing the jaguar cub at play, or the intellectual satisfaction of knowing how water moves through a rain forest, just as essential to the soul as food was to the stomach? To contemplate things we don't understand, to conserve that which has no "use," were uniquely human traits. If the rain forest and its complexity disappeared, we'd be impoverished intellectually and spiritually, as well as materially.

Whether the jaguar and the grison and the parrots would persist was the key question, but living in close proximity to such splendors filled me with gratitude. I felt the same way about Siberian tigers and Hawaiian monk seals, though I'd never seen them. In short, a world with jaguars and grison and tigers and monk seals was richer than a world without them. It was another selfish argument, to be sure, but it was a plea for conservation, and the basic research that made it possible, that anyone could understand.

16

Tropical Derelict

THE DAILY GRIND was getting to Chrissy. Her feet were sore and she was chronically sleep deprived. One evening, midway through the rainy season, she appeared at dinner sallow and deflated. We made eye contact over the rice and beans; she shook her head no.

She had risen at 4:15 to meet her monkeys on Standley, where they had gone to sleep the night before. She got one sample and then fell asleep, head in her hands, on a rock. Her beeping watch woke her, but the troop was gone. She spent an hour walking and calling, to no avail. By 3:00, after ten hours on the job, she was back in her lab, feeling like a failure.

It took Chrissy two days to find the troop after the debacle on Standley. But once she knew where they were, she went out with her video camera. Her reward was unobstructed access to fifteen minutes of monkey copulation. "Woo-hoo!"

A few weeks later she caught two more monkeys in the act. Chrissy came to dinner elated. "You should have been there," she scolded me, eyes beaming. "It was such a great day. I got all my samples early, and the monkeys were low to the ground all day."

She cut up the iceberg lettuce on her plate. "And," she said, looking around, "Leona sat on a branch about three meters off the ground and started reaching behind her to touch Robby. I was like, 'What's she doing?' " She took a mouthful of lettuce. "Then I see his erection. She was fondling his testicles! She was calling all the shots, but he wasn't exactly running away. Finally they start doing it. Twenty-one minutes!"

In most primates, copulation has been co-opted for social purposes — that is, they have sex not just to make babies but also to greet each other, smooth over differences, and form alliances. In some species, like bonobos, sexual interactions are nearly constant. But spider monkeys, Chrissy was finding, were different: they didn't copulate with great frequency, and their sex lasted so long — between fifteen and twenty-two minutes, in Chrissy's experience — that she wouldn't be surprised to discover the monkeys did it only when the female was in estrus.

While Chrissy took notes and counted the minutes, she could hear the rest of the troop calling to Leona and Robby, who remained silent. Afterward, they rubbed their bodies against branches, Chrissy said. "I think they were getting rid of the sex smell before rejoining the troop." Finished, the two monkeys raced toward the group, like kids who'd been absent without leave and suddenly felt guilty.

"Maybe that's why we didn't see Leona all day yesterday?" I suggested.

"Probably," she said, adding, "I thought she was pregnant."

"Couldn't you find out by analyzing a recent sample?" I asked. I was thinking about my own hormonal analysis.

Chrissy shook her head. "I don't want to know. It may change the way I see them." For several months I hadn't told anyone about my own pregnancy, for exactly the same reason.

Leona had just handed Chrissy a major datum toward understanding how estrus affects the social system of monkeys. She could now put a check mark in the column for "receptivity" — attempted copulations by a female. If Robby had been sniffing around Leona beforehand, that would have been evidence of her attractivity. Had Leona been leaving scent markings for days, that would be evidence of proceptivity.

Chrissy stayed fired up through dinner. It was days like this that reminded her why she was doing this work. "Who else here gets to come to dinner so excited and tell people what they've seen?" she said. "I'm out there all day, I see things." She was right. Before Chrissy had taken her accustomed seat, we were a subdued bunch. Three plant physiologists ate in silence, having spent the day on lab stools calculating rates of gas exchange. A couple of field assistants tried to see how many Saltines they could eat in exactly one minute. Even the researchers who'd spent the day in the field were hard-pressed to muster enthusiasm for storytelling. They were bone tired.

When people were tired or distracted or frustrated, it was only the one-upmanship of the more driven scientists — correcting someone's description of an animal, or topping the number of locations at which someone had done fieldwork — that made dinner conversation worthwhile. At times even these ploys fell flat. It all depended on who was at the table, their social skills and imagination, their energy level, and how long they'd been on the island. Sometimes it was just plain impossible to make conversation with people you ate with three times a day, seven days a week.

I joined Chrissy one morning, determined to get better at recognizing her female monkeys. At 7:30 we were in fine spirits. The air was still cool, and we were sweating only lightly as we climbed to the plateau, where the monkeys had been feeding lately.

At 8:30 we found the troop, lazily eating *Astrocaryum* fruits east of Standley Trail. Chrissy assigned me to Becka, who had dark eyebrows, pinkish, moon-shaped eyes, and a hook at the end of her clitoris. Chrissy had named her for one of her best friends in New Zealand.

My job was to watch and to wait. At 9:50 I collected my fecal sample — plus a few drops of urine on my head. The baptism, a first for me, brought me into good company. Up the River Teffe, an Amazon tributary, Henry Walter Bates, the British naturalist, came upon a tree full of cicadas. "On approaching a tree thus peopled," he wrote, "a number of little jets of a clear liquid would be seen squirted from aloft. I have often received the well-directed discharge full on my face; but the liquid is harmless, having a sweetish taste, and is ejected by the insect from the anus."

I shook my head, hard, and Chrissy laughed. "Get used to it," she said. We decided to perch on rocks that jutted from the steep hillside while the troop rested. Pippa and her infant, Lexy, sprawled on the horizontal branch of an enormous *ficus*, ten meters off the ground. Pippa was slumped into an extreme forward bend. Lexy lay with her long fingers resting lightly against her mother's back. Now and then she swiveled to follow, with her huge eyes, a passing butterfly or a falling leaf. I was riveted by the baby's interest in her environment.

A half-hour passed and the group moved on through the treetops, feeding here and there. As the monkeys hopped between a *Miconia* and a palm, Chrissy rattled off their names for me. "There's A.J., that's Scotty, here comes Becka. That's Di, with the fluffy face, and then Helena." She kept this up for the entire troop, twenty-one monkeys that seemed nearly indistinguishable except for the accident of their sex.

A *Callophyllum longifolia* rind dropped to the ground. Chrissy made a note. She wanted to know if nutritional factors were related to the ovarian cycle. Before her study began, she proposed asking whether monkeys suppress ovulation in times of food shortage. But since BCI was experiencing no dearth of fruit, she probably wouldn't be able to answer the question.

Even with binoculars, I found it damnably difficult to spot the monkeys. We stared, we moved a few feet, we stared. I began to feel dizzy. I found monkey-watching to be, at times, a stupefying business. When I told Chrissy my theory, that tilting the head limits oxygen flow to the brain, inducing sleepiness and vertigo, she just shrugged.

I crouched to stretch my back, and something fell to the forest floor. Immediately I was up: Who did that? I was outraged that a monkey would defecate before I'd identified her and readied a Ziploc.

There was nothing I could do about the sample near my feet, since I didn't know who it belonged to, and the troop soon scampered through the trees up a ravine. "You bastards! I hate you," Chrissy cried, struggling to keep up. They flew over an enormous patch of *Aechmea magdalenae* — a terrestrial bromeliad — but Chrissy and I had to go around it. Like a pineapple plant, the *Aechmea* had small spines along its swordlike leaves.

Down into another ravine we went, then up. Frank Chapman wrote that if BCI were somehow flattened out, with a cosmic roller, the island would double its size. Chrissy wasn't into the Stairmaster aspect of her fieldwork: she'd dropped two jeans sizes, which pleased her, but she was not terribly strong. "It's the smoking," she said.

Chrissy had been taking behavioral notes all day, but still she lacked Gracie's sample. At 1:05, Gracie touched her clitoris, then brought her hand to her nose. Chrissy noted the gesture: Was it sexual?

The relationship between hormones and behavior has long interested scientists and anyone else tantalized by the idea of biological explanations for human quirks. Biologists in Africa spent years correlating levels of testosterone in monkeys to their levels of aggression. The study had direct relevance to humans, as it was thought that the elevated testosterone levels of male prisoners explained, in part, their violent behavior. But it turned out that instead of the hormone increasing aggression, aggressive behavior elevated the level of the hormone. It was the violent atmosphere of prison itself that raised inmates' testosterone levels.

Here on BCI, Chrissy didn't care which came first, hormones or behavior. "I'm using the hormones only to find out when they're ovulating, and then to see if their behaviors are sexual. I want to see if hormones and behavior are correlated, not how one affects the other."

If you wanted to do that other sort of study, she added, "you'd have to perform ovariectomies, to take away the hormones. You'd see how their behavior is affected, and then you'd give them back the hormones, with pills, and see what happens." Even then, she continued, "the results on the captive population would be very different from the wild population. But you need both if you're going to make a comparison."

She interrupted herself now and looked toward the treetops. "Come on, Gracie," she shouted into the canopy. "Sit down and take a dump."

In a grove south of Zetek, we waited two hours for the blessed event, which happened about four times a day under normal fruit-eating conditions. In spider monkeys, digestion took from three to four hours; in howlers, which have the far longer intestines of leaf consumers, green matter that went in didn't reappear for more than thirty hours.

When Chrissy ran to retrieve Gracie's sample, she saw that it was packed with *Virola* seeds and nearly useless. Another problem was simply locating the tiny drops of fecal matter spattered among the leaves. If she was lucky, a sizable ball of the stuff would fall from the sky with a dung beetle riding on its back.

With their acute olfactory sense, these brown-shelled creatures of the family Scarabacidae arrived on the scene within a minute, like a deus ex machina to the needy primatologist. They came from up to a kilometer away. The beetles often found dung that Chrissy and I couldn't see, high on branches or leaves.

Using her front legs, the beetle worked the stuff into a ball, then rolled it backward while walking toward the leaf's edge. When the ball and beetle fell to earth, the dung beetle fortified herself with a few mouthfuls, then laid her eggs in the remaining ball so that the larvae had something to fatten upon when they emerged.

Dung beetles have drawn much scrutiny in the Neotropics because they're numerous and they cycle vast quantities of nutrients. In 1981, two researchers working in Panama baited traps with human feces and came up with twenty-two species of dung beetle in one location. Human scat, far more than that of any other rain forest mammal, is the type most avidly sought by dung beetles — in some locations, up to fifty different species have congregated on devoted researchers' samples — and the type most quickly sequestered.

In a Mexican rain forest, Alejandro Estrada and Rosamond Coates-Estrada examined the relationship between monkey droppings and dung beetles. The team placed twenty-gram piles of howler monkey dung, containing known seeds, along transects. Every thirty minutes, the researchers checked their dung piles — ten of them by day, ten of them by night — measuring how far the ball-rolling beetles rolled and how deep the burrowing beetles burrowed.

But beetles weren't the only creatures messing with seeds in their dung. The Estradas discovered that rodents could remove fifty-nine percent of seeds within twenty-four hours of the seeds' appearance on the forest floor — that is, if the rodents got there first. When dung beetles arrived first, the "faecal clumps" remained on the ground for an average of only 2.5 hours before being relocated and buried. The rodents, it turned out, weren't great at detecting seeds buried more than a few centimeters deep. To determine this, the Estradas had buried a variety of seeds in a wooden box at depths of between 0 and 12 centimeters, then put a mouse to work inside. (A control box contained marbles buried at the same depths. Given every opportunity, the mouse dug up no marbles.)

When the researchers were finished with their labors, it appeared that dung beetles were doing the trees an enormous service — and vice versa. Their study showed that one hundred percent of seeds dispersed by frugivores, such as monkeys, could be destroyed by rodents if they weren't first relocated by biotic and/or physical factors. Their study also found that howler monkeys, in that particular forest, dispersed the most seeds between April and October, which happened to be when dung beetles were most common in the forest.

The removal and burial of seeds by this abundance of dung beetles, they concluded, "significantly reduces the ability of rodents to locate the dispersed seeds." The beetles, by moving seeds horizontally across the forest floor and vertically down into the soil, also reduced clumping, which made seeds less detectable to seed predators and gave seedlings a better chance at survival.

By protecting seeds, and by recycling nutrients and energy in the ecosystem, the dung beetles were contributing to forest regeneration. Lifting their heads from their dung piles, the indefatigable researchers had come to see these winged coprophagites as "delicate ecological links relevant to our understanding of the self-sustaining capacity of the ecosystem."

A noble sentiment indeed, but if the dung happened to originate with Becka, Beri, Leona, Helena, or Gracie, and if Christina Campbell was around, the delicate link was, sadly, broken. The dung was now a fecal sample. "I'll take that; thank you very much," Chrissy would say, scraping the pellet into her Ziploc with a stick.

In midafternoon, three white-faced monkeys swung by our little gathering. A few male spider monkeys stood aggressively on their branches, but there was no real conflict. Chrissy was hunched over Gracie's sample on the ground, trying to separate fecal matter from seeds. All of a sudden she bleated, "Shit!" A white-faced monkey had defecated on her back. She stood to examine the damage: it lay in thick splotches on her shirt, her pants, her helmet.

"And it's got no seeds in it!" Chrissy cried. This was the quality of fecal matter we'd dreamed of all day long. A dung beetle fluttered around Chrissy's head. She bent forward to remove her helmet, then inadvertently flipped excrement from her ponytail directly inside her helmet. "So gross," she said, shaking her head.

Using tissues and water, we dabbed at the mess. Chrissy stank; the dung beetles wouldn't leave her alone. To make matters worse, by the time we pulled ourselves together, the spider monkeys had disappeared. Sometimes they crashed around, chattered, and threw down limbs as they traveled; other times, like now, they moved as silently as cats.

We listened and wandered and called for a dispirited half-hour, then headed wearily back toward the lab. It hadn't been a great day: Chrissy's binocular strap had broken, seeds riddled her fecal samples, of which she'd collected only four, and a white-faced monkey shat all over her.

As we trudged along, Chrissy sighed. "I could always open a pub," she said. "My boyfriend already makes beer, and I like to drink it." I nodded. "Probably make more money in a pub than academia anyway," she muttered.

The forest was unbearably hot now, and more humid than ever. When we worked in the field, our clothing stayed wet all day long. The hills were slippery with mud and the monkeys were hanging out in steep ravines, eating fruit with tiny seeds.

Following the troop in the rain was sometimes difficult. Over the patter of water striking leaves, Chrissy often couldn't hear the monkeys' movements through the trees or their vocalizations. When the rain got heavy, she stood under a palm frond and just waited.

In Tanzania, Jane Goodall, to reach her observation posts on the mountainside, had a long hike through "icy, water-drenched grass." To avoid spending the day in soaked clothing, she'd bundle the day's garments in plastic and hike through the early-morning darkness naked. For the first few days her body was crisscrossed with scratches from the saw-toothed grass, but then her skin hardened.

Exploring Africa, the Englishman Mansfield Parkyns wore a leopard skin over native dress, shaved his head except for a Muslim tuft, and coated his skin with cocoa butter. His solution to the wet-clothing problem was to remove all his garments, fold them neatly, and sit upon them until the sky cleared.

Halfway through Chrissy's field season, her new field assistant arrived. Kathy had responded to a notice Chrissy had posted on the Wisconsin Primate Research Center's online bulletin board. She'd worked with captive monkeys but had no experience in the field. On BCI, she took unhappy note of the weather and the biting insects, contracted a fungal infection, and within three weeks went home. "I guess I'm tougher than I thought I was," Chrissy said.

Another field assistant, Chrissy's fourth, was scheduled to arrive in just a few weeks. "This one climbed Kilimanjaro, did a month and a half on the Appalachian Trail, got malaria twice in Africa and still didn't go home," Chrissy reported. "So I've got high hopes for her."

It was the beginning of the end for Chrissy. A sturdy field assistant was lined up, fecal samples were accumulating, and behavioral data filled her computer files. I asked Chrissy if there were any trends or surprises when she looked at her numbers.

"No, not really," she said. "But I'm not looking for anything."

"You don't have some sense of anticipation about your study's end results?" I insisted.

"I really don't think about it."

This seemed odd to me, but I began to think of Chrissy's detachment as prophylactic. She didn't want to bias her observations, and thinking about her project's conclusion, with four months to go, would only frustrate her today.

So far, Chrissy had gathered more than 600 hours of group observations, including more than 45 hours of focals — each focal was five minutes of observation devoted to one individual. She would have liked to have more hours, but she didn't see how to manage it. There were still days when she couldn't find the troop, and there were days when she *wouldn't* find the troop, because she needed a day off. "I can't do this five days in a row," she said. "I've got bone bruises that won't heal, and there's no way I can work this hard without more sleep."

Studying spider monkeys at virtually any other site, Chrissy might have been better off. BCI's spider monkeys, the island's only troop, ranged over an area of 900 to 1,000 hectares. In Peru's Manu, two spider-monkey troops had home ranges of 153 and 231 hectares, respectively. The Manu monkeys daily trod a path that averaged less than two kilometers; on BCI, the monkeys' path was often three times as long.

Katie Milton, Chrissy's advisor, didn't care about these details. She thought Chrissy wasn't working hard enough. "Primates are very difficult to study, and spiders are hard. But there are other primates that run farther and faster than spiders, and people still get their data," she said. Milton spent some time studying spider monkeys on BCI, years ago, but it was during a season of torpor, when fruit was scarce, and the monkeys conserved energy by lounging in trees fairly close to the lab, scrounging food from the kitchen.

As Chrissy saw it, Milton had no conception of the job's physical demands. "You've just got to get better at finding them," Milton had said to Chrissy in her best Alabama drawl, smiling innocently. This kind of remark drove her nuts, Chrissy told me. When I suggested that she invite her advisor out for a day in the field, Chrissy said, unsmiling, no drawl, "I can't. She wouldn't make it, and I'd kill her."

Milton was right about one thing: the monkeys weren't Chrissy's only passion. Despite her long hours in the field and her early bedtime, Chrissy had managed to become the island's unofficial social coordinator. Her main qualification was a love of parties and expertise at delegating chores, whether moving the stereo to the old dining hall, icing beer, or mixing up two hundred Jell-O shots.

Though she complained about the burden, it was clear that Chrissy loved to be in the middle of things. It was her nature. Unlike her monkey troop, Chrissy had always been more fusion than fission, gravitating first toward an energetic, outspoken group and then quickly toward its red-hot center.

On BCI, Chrissy had been store jefe, video jefe, and change jefe. Now she'd risen to beer jefe, and part of the job was to decide when free beer would flow. BCI paid about twenty cents a bottle for beer and sold it for fifty. The profit went to a scholarship fund and also provided for free beer at parties and at certain Bambis.

Now Chrissy was planning another resident's going-away party. A group sat around the bar planning the drinks and talking about the Tropical Derelict Award. Hubi Herz was the current titleholder, which meant he was in possession of the Tropical Derelict T-shirt, a hole-pocked green rag signed by thirteen years' worth of derelicts. The T-shirt had never been washed, and it got worn by each derelict only twice: the night he or she received it and the night he or she passed it on. Emblazoned with a crest that read "Burnin' the Candle at Both Ends," the T-shirt was a high honor, awarded to the researcher deemed to both work and play with exceptional vigor. Chrissy coveted it.

BCI's early visitors would have been scandalized by such blatant disregard for appearances. In the 1920s, Alfred O. Gross, a Bowdoin College ornithologist, came to Panama to study the nesting habits of the purple gallinule and the Massena trogon. In addition to watching birds, Gross pondered the natural history of Panamanians. In a paper he suggested that it was the uniformity of temperature — to say nothing of days and nights of equal length and the constant march of budding, fruiting, leaf shedding, and breeding — that tired and sapped the energy of tropical denizens. If disease didn't get to the white men who visited the tropics, Gross worried, would lassitude?

David Fairchild's racism was more pointed. He noted the "poor intellectual showing of tropical peoples" and noted that there was a difference between visiting the tropics as a short-term visiting scientist and spending a lifetime there. "The latter carries with it all the dragging down of long association with peoples whose presence consumes one's attention but seldom stimulates thought; except as a race problem stimulates it."

Although Bates loved the tropics, he missed the intellectual stimulation of northern climes. He wrote, in *The Naturalist on the River Amazons*, "Humanity can reach an advanced state of culture only by battling with the inclemencies of nature in high latitudes." Thomas Belt, the English engineer and naturalist who traveled throughout the region in the mid-nineteenth century, lumped the human inhabitants of Central America with its plants and animals, which he believed were inferior to those of both North and South America and so lost out during the continents' biotic interchange.

Throughout most of BCI's history, it seemed almost a moral imperative for its visitors to appear energetic and productive. A prevailing Anglo idea held that life in the tropics made one soft, that it ruined work habits and morals. In photographs, Barro Colorado's earliest visitors — in their tight collars and ties — seemed pressed to make a counterpoint: that the exactitude characterizing good science had not, in Panama, been abandoned.

The Tropical Derelict Award did not appear until the mid-1970s, invented solely by gringos.

The going-away party was typical: dancing on the deck of the old dining hall, beer on ice, blender drinks, an elaborately decorated cake baked by Jenny Apple, who was studying the relationship between ants and caterpillars. Sometime after midnight, long after I'd gone to bed, residents offered their final farewells to the departing biologist and then presented the Tropical Derelict T-shirt to Chrissy. She was ecstatic.

"I wanted the recognition that I work and play hard," she told me the next day. "I don't think people think about you when you're not in sight, and I'm gone all day." I reassured Chrissy that it was obvious to anyone that she worked hard in the field. After particularly bad days, her feet dragged, her face was beet red, and her mood was often sour.

There was also the tenor of Chrissy's days off, which reinforced the impression of her days on — if only because she was so gratuitously slothful. Plotted, this off-day slothfulness rose in direct proportion to her work-day diligence. In my waterproof data book I noted Chrissy's behaviors: sleeping (till ten or so); traveling (toward the dining hall); feeding (omnivorously); traveling (toward her dorm); sleeping (for another two hours); traveling (to the lounge); and then resting again. Unperturbable, queenly, she'd inhale junk food and lie comatose on the couch for three videos in a row.

But Chrissy wanted the Derelict Award for another reason: "Now I'm part of BCI history," she said, grinning. Her name was added to the twenty-six others on the shirt. When she passed the shirt on, she'd add a personal artifact to the cardboard box that accompanied it, a box already bursting with a bottle of Shinola shoe polish, a twelve-ounce can of Spam, two strange-looking seeds, a candle, and a dozen tawdry-looking books, with titles like *Pleasure Ship Nurse* and *Bodies in Bedlam* ("Shell Scott, still up to his neck in guns, girls and gore"). Hubi contributed a bottle of banana wine to the box; Chrissy planned to donate a spider-monkey fecal sample.

The morning after Chrissy got her prize, after she had spun her favorite Abba CD and lip-synched to "Dancing Queen," Katie Milton sat at breakfast and drawled to a senior ant researcher, "Well, she got the Tropical Derelict Award. Now let's see if she gets a Ph.D."

*

One balmy morning when it happened not to be raining, I set out two hours behind Chrissy. If I found the monkeys, I'd find her. She was the subject of my focal.

I walked up Donato Trail, which was the trail closest to the spider monkeys' most recent sleeping site. The trail took its name from Donato Carillo, a campesino from Chirique Province who worked as Frank Chapman's assistant. Carillo had constructed the original stairway up from the boat dock — he'd stamped his signature on one of the last steps — and his trail started near the bottom of that stairway.

I soon came upon a pile of freshly gnawed *Spondias mombin*, which had just come into fruit. I didn't know if it had been howlers, white-faced, or spiders who'd been feeding.

I turned east, stopping now and then to listen. I knew that two troops of white-faced monkeys roamed this region. I heard nothing and headed south, contouring around some giant buttress roots and across a trickling creekbed. Eventually I heard a familiar chitter from the treetops. "Hello, monkeys," I said under my breath. I scanned the canopy but saw nothing.

Putting my hands to my face, I made my own sort of long call. "Chriiissssy!" No answer. I tromped on, looking for the familiar furry shape in likely branches. A fruit fell. A branch snapped and plummeted to the forest floor. After months of walking around the forest, looking at and listening to monkeys, I'd gotten a sense for how each of the three species moved about. Howlers rarely ventured below the canopy, and white-faced monkeys, who were much lighter than spiders, broke fewer branches when they leapt about.

I started searching the trees in earnest now, looking for a flash of red fur. Soon I was following two monkeys west, toward Donato, then I picked up two more. When I first went out with Chrissy, I could only react to her monkeys' movements. Now I tried to predict their route, based on past experience and on the forest. I knew where the *Virola* grew on this slope; I knew where the *Spondias* fell.

I called again for Chrissy but got no answer. If I had four monkeys, I thought, where was she? Then all of a sudden through the woods came an intense, repeated monkey scream. I dashed toward it, tripping on roots and sliding down the hillside. I heard Chrissy before I saw her.

"You got here ten minutes too late. You just missed Gracie and Robby copulating."

"Oh, hello."

"But I've got it all here" — she patted her video case — "on tape." Chrissy was red-faced, sweaty, and looking pleased with herself. Gracie was squatting on a low-hanging liana, her mouth a big circle that screamed at rhythmic intervals — a hoarse, raspy sound known as the warning call. What was the problem? Maybe she was upset about the group of white-faced monkeys that had come through before she and Robby began copulating, but now it was gone.

I watched Chrissy for a few minutes. She was laughing at the look of outrage still on Gracie's face. Robby lingered on a nearby branch, quiet and unimpressed. He moved a few feet away, then settled down again, gazing in turn at Gracie, Chrissy, and me.

"Did you hear me calling you?" I asked after a while.

"Yeah, but I couldn't answer. The tape was running."

During the copulation — as I would later see on the videotape — Libby, Becka's juvenile daughter, had sat on a nearby branch with her mother, who watched but did not interfere. Usually, spider monkeys copulated without any adults nearby, and never an adult male. Becka and Libby seemed watchful throughout the act, responding to any sudden noise, the cracking of a branch. They made no noise themselves and didn't respond to any long calls from the troop (or from me). When spider monkeys finish copulating, they usually rub their genitals on a branch and then hurry to join the troop. Today, though, Gracie continued with her warning calls, not eager to abandon her perch.

"I didn't even think Gracie was cycling," Chrissy said, pulling out her logbook. "I didn't think Jossie, her daughter, was old enough." The usual interbirth interval for spider monkeys was one and a half to two years. Now Chrissy was revising her estimate of Jossie's age.

"In the last week and a half, I've doubled the number of copulations I've seen." Chrissy grinned as she made a note. "That makes eight." A few days earlier it was Berri and Scotty, captured on videotape. The next day it was Berri and Scotty again. "Either the monkeys are doing it more or I'm getting better at reading their signals." Instead of switching off between females in a group, Chrissy was now sticking with just one, if she happened to be keeping to herself or if there was a male hanging around her.

Gracie quieted at last, then led Robby, Becka, young Libby, Chrissy, and me back up the slope toward Donato. There, the rest of the troop waited, gorging itself on *Spondias*.

I went out with Chrissy less and less frequently in the months to come, but we spoke often in the evenings about her work. It just went on and on: good days, bad days; days with samples, days without.

I'd leave BCI two months before Chrissy ushered her coolers full of fecal samples, labeled "BIOHAZARD," through the airports of Panama City, Houston, and San Francisco. The next time I checked in with her, she was back in California, running the samples through a complicated hormonal assay in a lab at U.C. Davis. The analysis would take eight months. She also had 85 hours of focals to correlate with the monkeys' daily reproductive status, and more than 1,200 hours of observations. When these were finished, she'd map the monkeys' behaviors — each keyed to a number — onto her hormonal graphs.

Nearly a year after Chrissy left Panama, she would get a lucky break. A BBC crew was filming BCI's spider monkeys and needed help in locating and identifying the members of the troop. It was a mutually beneficial arrangement: the BBC would finance her trip to Panama, and Chrissy would get to check up on her monkeys, to see which ones were pregnant and which had recently given birth.

The troop, by that time, contained twenty-seven monkeys. The six births confirmed the graphs that her hormonal assays suggested: cycling lines of progesterone and estrogen spiked at the time of conception and did not return to baseline.

Within a month of her return to the States, at the end of 1998, Chrissy dropped Katie Milton as her advisor and hooked up with Phyllis Dolhinow, a Berkeley anthropologist who specialized in primate social behavior. Chrissy presented a paper she'd written on spider-monkey fur rubbing at an American Society of Primatologists conference and won an award for the best student paper. After fourteen months on an insular jungle island, she was stepping out into the real world and doing just fine.

Life wasn't all rosy, though. As her work in the Davis lab drew to a close, Chrissy came to the realization that there were problems with her data. The field samples were "noisy": there was contamination from seeds and possibly from excess leaf fiber in the monkeys' diet, which decreased the amount of progesterone in the samples. And despite her best efforts, there was a chance that the identity of females had, at times, been mistaken, particularly near the beginning of the project. And then there was assay noise, from changes in temperature or lab protocols.

If Chrissy had more data points, more samples, the noise and the outliers wouldn't be a problem. But it was beginning to look as if Chrissy didn't have enough data. The hormones she was interested in cycled over a month, but daily nuances were far more important than she'd anticipated.

On BCI, Chrissy hadn't taken lightly those days when she couldn't find the monkeys, or the days she came in early from the field, exhausted. Now, in the lab, as Chrissy plotted data points along axes of hormones and dates, each missing fecal sample resounded ever more loudly. Chrissy had been right in the beginning: collecting daily samples from five female monkeys was hard.

What could she have done differently? Chrissy knew herself too well. The answer was, Not much. To collect samples from five study females in the wild, and to stay with each of them twelve hours a day, all she needed was two additional pairs of legs and two additional pairs of eyes.

"I had wanted my data to be perfect, and so I'm disappointed," Chrissy told me when we met for a beer at a Berkeley pub. "But this was my first time, and the monkeys move a lot. I should have done a basic behavioral study for my Ph.D. and saved the hormonal stuff for a postdoc."

She sipped her beer. "Now I'm thinking of this project as a pilot. I'm still interested in the same questions. And now I'm even more interested in dietary effects on their behavior and in pregnancy itself. If I do the study again, I'll have at least two field assistants. I need someone to stay with a female when she leaves the troop. Radio-collaring a female would be the way to do it, but the collars cost fifteen hundred dollars each. No one gives that kind of money to a doctoral candidate. But to a tenured professor, sure."

While Chrissy couldn't correlate behavior with hormones for every day she spent in the field, her data still told a good story. She could use them to compare cycling behavior with noncycling behavior, and the hormones of pregnancy with those of lactational amenorrhea, or breastfeeding. And she had proved that spider-monkey fecal samples could be used for hormonal analysis. That in itself was novel, and worthwhile.

"The data aren't as fantastic as I thought they would be," Chrissy said, toying with a ring on her finger. "But this is stuff no one's gotten before." She looked up for a moment. "Dissertations are never the beginning and end of a project," she said. "They always open up more questions."

17

A Maverick Reconsiders

LUTZ RAVINE, on the edge of the forest just south of the lab clearing, was a bat superhighway. One night Bret strung a three-meter mist net over Lutz Creek and another six nets along Donato Trail, which defined the ravine's southern edge. Every ten minutes we ran our headlamp beams the length of each net. If they came up empty, we sat down and waited atop a boulder in the creek.

Capturing bats required great patience, not so much for waiting out the flying mammals as for disentangling their heads, leaf noses, and fragile wings from the net filaments. Despite this impediment, Bret had five bats within an hour, each differing slightly in size and coloration, all of them leaf-nosed. He handled the animals deftly and with nonchalance, as he did just about any creature he came across. He shared this quality with many a naturalist.

In the Amazon, Henry Walter Bates had nearly constant encounters with all kinds of creatures, through which he maintained the stereotypical British stiff upper lip. In the Brazilian village of Caripi, he found himself in a room long unused:

> The first few nights I was much troubled by . . . vast hosts of bats sweeping about the room . . . After I had laid about well with a stick for a few minutes they disappeared amongst the

tiles, but when all was still again they returned, and once more extinguished the light. I took no further notice of them, and went to sleep. The next night several got into my hammock; I seized them as they were crawling over me, and dashed them against the wall. The next morning I found a wound, evidently caused by a bat, on my hip. This was rather unpleasant, so I set to work with the negroes, and tried to exterminate them. I shot a great many as they hung from the rafters.

Identifying the dead, Bates realized he'd slept with four different species, belonging to three genera: *Dysopes*, *Phyllostoma*, and *Glossophaga*. "According to the negroes, the Phyllostoma is the only kind which attacks man," Bates wrote. "Those which I caught crawling over me were Dysopes, and I am inclined to think many different kinds of bats have this propensity."

Historically, there has been a great deal of confusion about what is and what isn't a vampire bat. The actual blood feeders — three closely related species that make up the phyllostomid subfamily Desmodontinae — don't even have vampire in their name. Yet several genera of fruit eaters had retained designations like *Vampressa* and *Vampyrodes*. A nice reminder, Bret thought, that unchecked assumptions can be dangerous.

"The Phyllostomidea contains 143 species," Bret told me as he sat in Lutz Creek and stroked a bat's fur. "But only the three desmodontines drink blood, and only one of the three species normally drinks the blood of mammals. After long debate, it seems clear that this subfamily evolved from ancestors that had fleshy nose leaves, but they lost them as they developed the blood-feeding habit. It's really a marvelous testament to the effectiveness of natural selection."

He held up the bat to examine its twitching nasal megaphone. "Nose leaves are thought to focus echolocation calls while the animal is holding a large item in its mouth. But vampires carry blood in their bellies, so their mouths are never blocked. Somehow, the cost of having a nose leaf was enough to favor smaller and smaller nose leaves, until the leaf was completely eliminated." He lowered the bat, still stroking her.

"Every time I think about it, I'm amazed all over again." Now he placed the bat in a small cloth bag. "There you go, sweetie, it's just for a minute."

Bret released two of his five bats because they weren't his study species. Then we slid the nets closed on their supporting poles, collected our gear, and hiked back to the lab. Passing the lake edge, we stopped to sweep our headlamps over a shallow cove. Six yellow orbs, from three caimans, glowed back at us.

Up at the bar, Bret extracted another bat, also named Sweetie, from her bag. From the shaved patch on her back, we could tell that she'd recently worn his transmitter. With a hand scale, Bret weighed the cloth bag empty and then weighed it again full. The bat weighed nine and a half grams. When Bret affixed the transmitter, over a week ago, the bat had weighed ten grams.

"Good," Bret said. "Now when someone asks me at a seminar how much the transmitter stresses them, I can say just a little. Also, this animal probably hasn't eaten yet tonight."

He fed the bat some sugar water from a small bottle and returned her to the bag. Next, he trimmed some fur from the back of one of the remaining two bats. He noted its species — *Artibeus watsoni* — in the logbook, and added the bat's sex and weight, the date, the time, and the frequency number of the transmitter. These transmitters cost $200 each; stored inactivated in a refrigerator, they kept their charge for about six months. A transmitter usually stayed on a bat for a week before falling to the forest floor, where it would beep — for no longer than two weeks — until Bret retrieved it.

To the trimmed area Bret glued the transmitter. He dusted the area with flour, to keep debris from sticking to any stray glue drops, dripped a little sugar water into the bat's mouth, and placed her back in the bag. The entire operation took about ten minutes.

By now, several residents had gathered in the bar and it was no longer the peaceful sort of place preferred for manipulating wild animals. A German bat researcher named Dietrich arrived with a recently netted *Artibeus jamaicensis* in a bag. The crowd clamored to see it and, of course, take its picture.

Dietrich held the bat with wings spread, nearly thirty centimeters across, against its burlap sack. Displeased, the animal opened its mouth and bared a formidable row of teeth. Hubi clapped his hands to his ears. "Aggh!" he cried. "Doesn't that hurt your ears? It's killing me." Everyone laughed. The cries of *Artibeus jamaicensis* are ultrasonic.

Bret had one bat left, an animal he'd tagged before. But it was no ordinary *watsoni* or *phaeotis*. In fact, Bret didn't know its species. Scanning a bat key, he ran down a list of questions. Are the stripes on its head distinct or faded? Are its eyes small or medium? Are its molars sharp?

"Its teeth are too small for me to see even with a hand lens, so that's irrelevant," Bret said. He found a lot of these questions inane; without a sample bat to calibrate against, the questions were subject to so much interpretation. And none helped him determine the bat's species.

Bret had previously tracked this animal, a female, to a roosting spot in an unmodified leaf. That was a clue: the leaf told him the bat was neither *watsoni* nor *phaeotis*, since both always modify their roosts. "This could be a hybrid," Bret said, scrutinizing the edges of the bat's ears, which would be yellow-rimmed on a *phaeotis*, cream-colored on a *watsoni*.

That he might have a hybrid in his hand didn't impress Bret. The traditional view of animals held that independent evolutionary pathways stuck to themselves. But now that DNA analysis is routine, it appears that animals' gene pools are far less rigid than anyone in the premolecular age ever expected, and that hybridization among higher animals — especially in younger lineages — is hardly rare. (Plants intermix all the time; it's thought that as many as forty percent of plant species arose through hybridization.)

"As biologists, we're interested in species identity," Bret said. "But lineages are interested in persisting. They don't have an allegiance to particular niches, Latin names, or historical connections."

A standard species definition for animals uses an interbreeding criterion: if two animals can produce fertile offspring, then they're the same species. If the parents of the bat in Bret's hand were *Artibeus watsoni* and *Artibeus phaeotis*, and if this bat was fertile, then *watsoni* and *phaeotis*, using this definition, were the same species.

Bret was not a big fan of the species concept. "It's a necessary evil," he said. "We're forced to define the term for all kinds of practical reasons, but there's a very good reason that we scientists can't agree on a single definition: Nature has not cleanly defined the concept herself."

The power of intercrossing lineages, Darwin wrote, "is not hypothetical." Because such crosses produce novel combinations of genes — about which Darwin knew nothing, though he suspected that something like them acted as a source of variation — they may even be responsible for the evolution of new species.

"Successful leakage of genes from one species into another," wrote the evolutionist Ernst Mayr, is "a self-accelerating process." Should successive generations of hybrids mate with each other, the process speeds up, "until ultimately the two species are connected by a continuous hybrid swarm."

It was impossible to know if Bret's bat was the harbinger of a new species. It takes generation after generation to raise unbreachable barriers between species, and when a "new" species clambers up that barrier, all manner of changing environmental conditions could knock it off, sending it tumbling back toward the friendly confines of either *watsoni* or *phaeotis*.

In affixing a transmitter to the mystery bat, Bret hoped to learn more about its habits and thus refine his identification. *Watsoni* and *phaeotis* make their living the same way, searching for fruits and eating them in temporary leaf roosts. But probably to avoid directly competing with each other, the bats roost in different plant species. It is a perfect example of niche partitioning, in which closely related species alter their behavior to avoid direct competition for the same resource.

Bret dusted the area around the transmitter with flour, gave the bat some sugar water, and gently placed her in the third bag. The bar was clearing out now as researchers drifted back to their labs. After collecting the equipment, we returned to Lutz Creek with the three bats. Bret released two and handed the third one to me. She rested on my open palm for two long seconds, then realized she was free. I felt her pointy little claws as they left my hand, then just a warm puff of air. She swooped over the creek and disappeared into the night.

When I first started hanging out with Bret, he told me that studying bats was, for him, a way to get at far bigger questions. He didn't elaborate at the time, but as the months passed, it became clear that he was less and less content simply to work his niche among Neotropical bat researchers.

It wasn't just the obstacles to getting good data; Bret had enough bat material to fashion a few chapters of his dissertation. But bigger ideas preoccupied him. As a young grad student he'd read a paper by the evolutionist George Williams on the tradeoff between youthful vigor and eventual senescence. He found the concept interesting — "The tradeoff concept had real power," he said — but he filed it away. Over time, though, Bret began to see patterns in nature that fit the evolutionist's tradeoff paradigm. Had Williams hit upon a general principle?

In addition to Williams, Bret was heavily influenced by the late William D. Hamilton, who is widely considered to be the greatest evolutionist since Charles Darwin. Hamilton, like Darwin but unlike so many biologists, was a keen naturalist. The thousands and thousands of observations in his head allowed him mentally to evaluate and test his ideas. Bret, walking through BCI's forest, began to accumulate his own store of observations against which to test his own developing ideas. Could the latitudinal diversity gradient, Bret wondered, be explained through the concept of tradeoffs?

Bret, who'd never been particularly interested in narrow specialization, yearned to tackle big questions. "And this was a gem," Bret said, in a gee-whiz tone of voice. We were sitting in the bar on a Thursday afternoon, a bag of yucca chips between us. "I thought the explanation had to be big and obvious. Anyone could see the pattern, but no one had satisfactorily explained it. I figured they must be overlooking something."

On BCI, Bret started to look. He had plenty of time to think and to stare into the canopy. Over rice and beans, he shared his ideas with visiting scientists. He talked with Alan Herre about fig wasps, with Lissy Coley about the allocation of chemical defenses in plants, with Robert Stallard about processes upon which natural selection had no effect. "Eventually I realized that everything I said about adaptation could be profitably rephrased within the context of tradeoffs," Bret said.

Tradeoffs are constraints that limit organisms from succeeding in all places at all times — benefits in one process bought at some expense in another. Examples of tradeoffs in nature abound. A tree can't grow quickly *and* have dense trunk wood. A plant can't be heavily defended against herbivores and pathogens *and* grow rapidly. Predatory robberflies raise their body temperature by basking in light gaps, which allows quicker takeoff and agile flight. But the physiology that makes the animal more effective in light gaps makes it more torpid, and less effective, in the shade.

Over the course of several months on BCI and many more months back home, Bret would begin to fashion an explanation for the latitudinal diversity gradient that was based on two universal tradeoff relationships: that between robustness and energetic efficiency, and that between niche breadth and niche efficiency. His argument began with an observation: while tropical species are clearly excluded from the temperate zone by the climate, many temperate species grow well in tropical gardens. It is fierce competition from local species that excludes temperate species from tropical habitats.

Temperate species persist through vast annual temperature fluctuations and, on longer time scales, through glacial and interglacial periods. The ones that succeed are those that allocate substantial resources to withstanding these climatic extremes. That is, their gene pool has to contain information that will help them survive the coldest winter and the hottest summer; between extremes, they maintain a lot of useless machinery. But temperate species also have to compete with one another for so-called limiting resources — light and nutrients, for example. Unfortunately, these two selective factors act in opposite directions on species robustness.

"There is no stable balance for selection to find," Bret explained. "Temperate species, then, are jacks of all trades, masters of none. They can't optimize, and so they are inherently prone to go extinct at the hands of both antagonistic climatic forces" — droughts, for example — "and competition from similar species."

Tropical species, living under relatively stable abiotic conditions, have no need of broad tolerances. Instead, they have the luxury of allocating their resources to a specialty, such as capturing light in a wet tree gap. Tropical species, facing less disparate extremes, tend to exert constant directional selection pressure upon one another. The result is many optimized species competing for limiting resources.

Bret continued. "As temporal fluctuation decreases" — that is, closer to the equator — "optimizing selection should tend to fragment an inefficient generalist species into a number of more efficient specialists."

But how many specialists can there be? How many microhabitats do the tropics contain? There aren't, in fact, that many ways for such organisms to be different. Numerous species, Bret believed, will tend to find the same shared limits to optimization. At this point, species will be "approximately equal competitors with little power to competitively displace each other."

Here, Bret's theory abuts the interspecific competition hypothesis, which argues that competition among species for limited resources results in increased specialization. These highly specialized species pack into extremely narrow ecological niches, which reduces competition among them for their exclusive resources.

I went inside to top off my water bottle, and Bret fetched more yucca chips from the store, which was now sharing lab space with Katrin and several buckets of arthropods, collected from the *Atta* dumps. When we reconvened, Bret offered me an analogy.

"Imagine two travelers in the temperate zone. One dresses in heavy winter clothing and carries a sleeping bag, tent, and stove. The other wears running shorts and carries no gear. In the summer, the light traveler will get to food sources far quicker than his encumbered competitor, and he'll possibly spread around more genes as a result. But come winter, the gear-laden competitor will easily outpace his scantily clad cohort — who may even, by now, be dead."

Put the two competitors in a warm, stable climate, he continued, and the speed specialist will crush the ponderous, ready-for-anything generalist. Unless, of course, the generalist finds a way to shed his unnecessary gear or exploit some empty niche with no competitors in it. In other words, unless the generalist learns to specialize.

"In short," Bret said, "it's the tradeoff gradient between tolerances for these pressures that predicts the relative diversity of different habitats."

Contrary to the competitive exclusion principle, which states that two species cannot coexist indefinitely if they are limited by the same resource, Bret — borrowing from Robin Foster and Stephen Hubbell's community-drift model, derived from study of their fifty-hectare plot — held that not only could equal competitors coexist, but that competition itself produced such equal competitors. Species were failing to compete each other into extinction in the tropics; moreover, species were actually accumulating, due to immigration, drift, and high speciation rates.

What Bret saw when he looked around Barro Colorado were unique resources that had, a long time ago, been adopted by one lineage or another. Once the environment was competitively saturated, pressure to find a unique specialty decreased, in a relative sense, because all potential niches had been filled. At this point, the only safe haven from competitive exclusion was to become an extremely efficient competitor in an already inhabited niche. That meant becoming similar to other organisms. Trees, for example, didn't find a new specialty; they found the same specialty. And that, according to Bret, was why so many tropical trees looked alike.

"It's a bit like a competitive market," he said. "When a market is new, there's lots of room for innovation. In that early stage, competition generally takes the form of filling new niches with products that do something novel. But once variety has reached some level, variations between niches become too small to constitute a basis for distinction. Then competition starts to reward products that do essentially the same thing faster, cheaper, or better. Each product exerts competitive pressure on all the others until all the surviving products bump up against some external limit, rendering all the survivors equal."

Bret paused while I digested his analogy. Unable to resist, he added, "Then, you call in the advertising guys to convince customers that your product will make them more attractive to potential mates."

Tradeoffs have been part of the scientific conversation at least since Darwin's time, but where most scientists have considered tradeoffs incidental to organismal design, Bret considers them organizational.

"There's a battle between those who think adaptation is utterly common and those who think natural selection is generally weak, and that adaptation is therefore rare," he said. Some evolutionists don't seem to believe in adaptation at all: they believe that organismal design is dominated by failures of selection to find possible solutions. Bret also believes in constraint, but of a very different kind — a constraint that arises as a byproduct of structural necessities. To those who view selection as weak, random genetic drift somehow explains complex characteristics, such as behaviors, like migration, and complicated morphological features, like nose leaves.

Bret believes that drift is much more common than selection, but that it rarely explains anything of interest. "To me, God is a better explanation than drift," he said. "And adaptation, constrained by tradeoffs, is a much more powerful explanation than God." Since it is extremely difficult to run controlled experiments to answer questions about adaptation and drift, the best way to examine the forces of evolution, in Bret's opinion, is to study tradeoffs as they manifest across similar species.

"Once I realized that what typically limits adaptation is tradeoffs pulling in opposite directions, everything clicked," Bret said. "Every question had more depth. It wasn't 'Why are bat tents adaptive?' but 'What are the costs and benefits of roosting in and making tents?' "

As his time on BCI drew to a close and his thinking on the latitudinal diversity gradient began to coalesce, Bret turned his attention to another question that he'd been mulling over since the early stages of his graduate career, a question he formulated after attending a presentation on telomeres, a cellular structure. Convinced of the power of tradeoffs, he asked what implications evolutionary theory had for medical science. Specifically, he was interested in a possible link between cancer and senescence.

With just a few months left on the island, Bret now spent

the majority of his time massaging his tradeoff theory and searching the scientific literature for evidence to support his telomere theory. Together, these ideas would form two-thirds of his thesis. Tent-making bats, which had drawn him to Panama in the first place, would fill out the remainder.

It would take Bret another year to get all his thoughts down on paper formally, but here in Panama he was laying the groundwork, talking through his theories with visiting scientists and banging out the roughest of dissertation drafts. Now and then, he tried out his ideas on me.

"How do you make a complicated, multicellular organism without tumorous growths?" Bret asked me late one afternoon as we sat in the bar again. "In a body with ten trillion cells," he continued, "there are ten trillion opportunities for runaway cell lines to scrap the entire project. So how do you build a system, a fail-safe, to keep the renegades from overrunning the place?"

The renegades were cancer cells. And to Bret's way of thinking, evolution had already supplied vertebrates with just such a fail-safe. And true to form, it was a tradeoff. "Cancer and senescence are somatic opposites," Bret said. "One is runaway cell growth, the other is retarded cell growth."

Our cells, Bret explained, have finite proliferative capacities, called Hayflick limits, that cut off the unregulated growth of any cell line, thus preventing tumors. But that same finite proliferative capacity also limits tissue repair and replacement, causing gradual degradation of somatic function with age. And so we senesce.

"There's a tradeoff," he said, "between having an increased capacity to repair damage and living tumor-free." Bret argued that the traits, which evolved together, could not be uncoupled. Natural selection had already optimized both our longevity and our capacity to live nearly tumor-free during our peak reproductive years.

Bret's thinking derived from the Williams paper that had inspired him as a young graduate student. The paper stated that natural selection opposed senescence, because a longer life would have more reproductive opportunities than a shorter one. But in the absence of senescence, lifetimes would still be finite: people would die in accidents, they'd starve, they'd be killed. Given that, Williams argued that natural selection should tend to favor early reproductive opportunities. And so traits that have beneficial effects early in life, at a cost paid later in life, will tend to spread. "Individuals are thus endowed with youthful vigor," Bret would write, "a trait for which they pay an ever increasing price."

Bret's theory began with the telomere mechanism. Telomeres are long, repetitive sequences of DNA at the end of chromosomes. They don't make proteins but instead protect the ends of the chromosomes so that the failure of polymerase enzymes to copy the very tip of the chromosome doesn't begin to destroy genes. Telomeres act as a buffer for important genes that would otherwise be lost.

One downside to this rather neat system is that our capacity for self-repair is not infinite. It's now believed that the erosion of telomeres below a critical length triggers "cellular senescence," the shutdown of the machinery that allows cells to replace themselves. The enzyme telomerase, which is capable of elongating telomeres, is present in fetal cells and sperm-producing cells. It is also expressed in a diverse array of tumors, but it is curiously absent in most normal adult cells.

The whole system, according to Bret's theory, is an antagonistic pleiotropy — wherein one gene has a beneficial phenotypic effect early in life at some cost later in life. In this case, eroding telomeres prevented tumors from erupting all over the body, but they also placed a limit on the extent to which the body could maintain its tissues.

In an abstract for a journal paper based on his thesis, Bret would soon write, "It may seem we are a taxon plagued by senescence and tumors, but in fact we are the beneficiaries of extensive, vigorous lives that result from selection's remarkable efficiency at simultaneously minimizing the harm of these two opposing hazards."

Where science saw the solution to the cancer problem as a problem in itself, Bret saw the solution — the erosion of telomeres — as an elegant evolutionary innovation. "The evolution of the telomerase system was no serendipity," he told me, leaning forward in his plastic chair. "If there were some simple way to modify the parameters of that system, to further extend life without significant costs, selection would have found it." In short: maximum longevity can't be greatly extended without a dramatic increase in the rate of tumor formation, and increasing tumor suppression would accelerate aging.

If Bret was right, then seeking simultaneous cures for cancer and senescence in telomere research was folly. "We can probably slow the aging process if we're willing to endure an increased risk of cancer," he said, "but we won't evade the tradeoff." Bret read to me from a rough draft of a chapter: "Diseases that kill millions of people every year cannot be cured simply by switching a gene on or off. If it were so simple, then natural selection would have done it long before we ever became aware that we had a problem."

Bret smiled. I smiled. It seemed so simple, so elegant. But was this any way to get a Ph.D.? What about all the hours in the forest listening to beeps and filming leaves? "I'm still writing about bats," Bret said, settling back in his chair, "but in the context of tradeoffs. The most important part of my dissertation will be that central concept. The diversity gradient, cancer and aging and tent-making serve as examples of something larger. What I'm doing is demonstrating an ability to generate hypotheses and then attempting to falsify them. Contrary to the conventional wisdom, falsification doesn't have to occur in the forest or the lab. It can also be done in the library."

Near the raft Gloria lingered. A gecko scooted across the underside of the balcony roof, startling me. Briefly, I felt as if I were living in a zoo where all the cages were open and the keepers had run away. Basilisks lolled in the mud near my dorm, and white-faced monkeys hung out in the nearby balsa trees. *Bufo marinus* — for its apparent stupidity the most maligned animal on the island — squatted on the walkways. The hallway of my lab was visited freely by katydids, mantids, crickets, and geckos. Sometimes a bat found its way in. One night, it was a plague of green lizards. On the walkway between the labs I came upon *Megasoma elephas*, the rhinoceros beetle. Related to scarab beetles, it has one long horn that curves up toward its forehead and a shorter one that curves down over its eyes. Its body is the size of a grenade, and except for its hairy legs it resembles a tank. Prized by collectors and dependent on mature, lowland forest, *Megasoma* is now considered rare.

My attention wandered back to Bret. If he was right about cancer and senescence, then biotech companies, med schools, and pharmaceutical giants were wasting a great deal of time and money trying to make cancer cells mortal and healthy cells immortal. It was a bold accusation from someone without an advanced degree, from someone who had, so far, studied only bats, not cancer or telomeres or even the human body.

Bret expected that his ideas might win him accolades in some quarters, but he also expected that his thesis, which he hoped to publish, would piss a lot of people off. At the very least, Bret hoped that his work would introduce a new vocabulary to medical science and bring together the far-flung worlds of gerontology and evolutionary theory. If his hypotheses panned out — his thesis would suggest numerous ways for medical practitioners to falsify his claims — it could lead to medical benefits.

"For example, telomerase treatment, in vitro, may rejuvenate tissues or organs before transplant, extending telomeres in accordance with the amount of cell division expected to occur in the recipient," he would write. Bret thought there might also be a useful HIV therapy in telomerase.

The point, he said now as Gloria dove, was that practitioners could use telomere therapy in research areas that natural selection doesn't reach, as had happened in the cases of surgery and of in vitro methodologies. "Given our increasing ability to detect and surgically eliminate tumors, we may one day be willing to accept an increase in our tumor risk in order to extend youth," he'd write. "The in-vitro lengthening of zygote telomeres would likely produce that heritable effect."

Bret leaned back, put his feet on the table and sighed. "Writing a thesis based on theory is risky," he said. "I could work for years on these ideas and produce nothing."

The reality check surprised me; Bret had sounded so confident. But he wouldn't know until he tried to get his paper published whether the scientific community thought he had something worthwhile to say. A long moment passed, and then Bret's doubts went back into remission. Quietly but insistently he said, "I think this is going to be big."

One afternoon Bret and I climbed to the top of the canopy tower and watched a lightning storm move north out of Gamboa. Bret's tenure on the island was almost up, and he was becoming increasingly philosophical about his field period.

"I accomplished some of what I wanted to in the last fifteen months," he said, pushing his hair behind his ear. "I learned a lot about tent-making bats, and they're doing what I hypothesized they were, even though I may not have a robust enough data set to convince others of that."

He took a sip from his water bottle. "The messy part of science is that you don't nail shit down. I guessed at first that small bats built tents when flying was expensive. I predicted two levels of ranges: a small one for nightly foraging, and a larger area for locating new food patches. And this is what my radiotracking showed: these are very local bats, and they don't fly much unless they're relocating. For them, tent making is a worthy tradeoff."

Unfortunately, though, Bret's relocating data were slim, because the radiotransmitters didn't stay on long enough and put out only a faint signal. One day Bret went searching for a fallen transmitter. He was boating around the island's shoreline when he picked up the signal at the end of Barbour Trail, 2.3 kilometers from Lutz, where the bat had been foraging. "That was my shining gold piece of data," Bret said. "It showed I might be right." The dropped transmitter indicated that the bat flew longer distances only when it needed to find a new resource patch. And the transmitter had landed near an actively fruiting tree, suggesting the likely motive for the move.

We turned our heads now at the sound of a snapping branch, but we saw nothing. "The tropical forest is the most complicated thing in the known universe, and it's hard to figure out what's true from what looks to be true," Bret continued. "In physics, the data are unequivocal. When you're right, you know it. In the forest, that's hard. You have to train yourself to figure out what's right.

"On BCI, I had the luxury of thinking and of high-quality people to interact with. STRI, in fact, has been more of a university to me than my real university."

Obviously Bret had learned a great deal from his tropical adventure, both within the standard curriculum and along its entangled banks. His broad-ranging experiences had fed his wandering mind, leading him from tent making in bats to design-constraint tradeoffs to telomere theory. Ultimately, he had worked around constraints both technical and personal to approach what was, for him, a more optimal intellectual challenge.

Heavy purple clouds were sweeping faster up the canal now, but we made no move to leave the tower. "Before I started working on bats here," Bret continued, "I didn't have a good appreciation for the way science is done. Now, I'm sorry to know."

I was taken aback. "What do you mean?"

"At some point in your study, you know the answer to your

question — you satisfy your childlike wonder — but you can't demonstrate it. Sometimes in science you get a huge data set and you don't know what it means until you sit down and analyze it with a computer. But with small sample sizes and dramatic events, like my bat project, it's easier to get a feel for what's going on."

He looked off across the low cordillera and continued. "If only my animals *flew* farther, the gruntwork would have been less gruntlike."

"You were here for a long time," I said.

"Yeah," he said. "But fifteen months on BCI was good for my mental health. Now I'm ready to look my department in the eye and say, 'You were wrong. I *was* ready to do my thesis research.' "

There was an unfamiliar set to Bret's jaw. I had the feeling he was practicing a defense he'd soon need to employ.

The storm was upon us now, and I screwed the lid onto my water bottle. Bret tightened the lace on his boot, and together we clambered down from the tower as the first drops of rain hit the canopy.

18

The Great Unknown

EVERY YEAR, Robert Dudley, an entomology professor from the University of Texas, made a pilgrimage to Panama to census migrating Lepidoptera, to study frog vocalizations, and to get his annual tropical fix. BCI was Robert's favorite place in the world: he dreamed of building a little house just beyond the last dorm and setting up shop — reading, researching, writing papers, and observing the island's higher primates.

Robert was something of a polymath; he spoke four languages, traveled incessantly, and read widely outside his field, which was unusual for a scientist. He wrote papers on the evolutionary origins of human alcoholism, Antarctic mollusks, and rattlesnake strikes, and he was a leading authority on insect biomechanics. He also spent a fair amount of time thinking of ways to murder people on BCI.

"This would be the perfect place," he'd say, strangely insistent. "Think of all the ways you could get rid of the body." There was the lake. There were the carnivorous animals in the forest, the insects and fungi that made recently living organisms disappear virtually overnight. Phone lines could be cut, boat engines disabled. Living residents often disappeared for days on end — in the forest, on vacation to the Caribbean coast — so the postmortem interval would be roomy. Camouflage and crypsis, two mainstays of survival in the forest, contributed to the psychological ambience: so many things were not what they appeared to be.

There had already been one murder on BCI. In 1968, a kitchen worker, heeding his insidious inner voices, stalked a Colombian named Ruiz to the monkey cages in Allee Creek and stabbed him to death. Declared insane, the murderer, a man named Moore, was incarcerated in a Panamanian institution, from which he escaped several times. On one such foray he took a job at the Summit Golf Course, outside Gamboa. There, directed by the same inner voices, he murdered a caddy. Having been delivered to the mental ward of Gorgas Memorial Hospital, he soon committed suicide.

If Robert heard voices, they surely spoke in the same scientific idiom that he did. Even when he talked about personal matters, his speech was peppered with words like differential, inverse relationship, order of magnitude, subset, correlate, and constraint. Of matters scientific or not, he spoke with a wonderful precision. His highest praise, in fact, was to say "Precisely" to anything he agreed with.

Another of Robert's eccentricities was his daily jog along the forest paths or down creekbeds. No one else on the island, except sometimes Hubi, would actually run through the forest — at least not without some urgent purpose. But Robert found it invigorating.

As he ran, he listened. "That's a trogon," he'd say, cocking his ear. "That's a motmot." Robert was good with birds, and with insects and frogs. But he didn't know much about the larger vertebrates, animals most people consider familiar and therefore easier. I once showed him the jawbone of a coyote, part of my own collection, and he had no idea what animal it came from.

For a few weeks in September I went out every morning with Robert and drove a boat around the lake, at top speed, in pursuit of *Urania fulgens*, a migrating diurnal moth. Robert stood in the bow, precarious in his Tevas, with an anemometer in hand. I'd get us within a meter of the flapping creature and he'd lean out over the water to take readings of its flight speed, measured in meters per second. Robert was exhilarated by the moths, calling out, in his slightly nasal voice, "Fantastic, fantastic!" as we chased them across Gatun Lake.

I was exhilarated by getting to drive so fast and so erratically. Many times Robert nearly fell out of the boat when we suddenly changed course. His shins were bruised from falling against the metal gunwales. He didn't seem to mind — it was part of the job.

The point of all this chasing and measuring was the investigation of the aerodynamics of flapping locomotion and the amount of energy it took to sustain the erratic flight of moths. At other sites, Robert studied hummingbird flight and gliding lizards, but Lepidoptera were great, he said, because they were conspicuous and easy to observe.

The day began with a census: we'd tie the boat to a stump near Buena Vista Point and count passing *Urania*, in two directions, for exactly one minute. Some days only one or two moths crossed our imaginary tripwire; some days it was fifty. *Urania*'s migratory range was believed to be several thousand kilometers, some of it over open water. The moth migrated annually, but unpredictably, through central Panama, but its intensity was highly variable. In some years, observers counted a couple of hundred moths; in other years, it was hundreds of millions.

Once Robert got his flight-speed reading, he began swiping madly at the flapping moths as I gunned the throttle. He was good with the net, but it was even more precarious work than the anemometer. He parried, he jabbed, he leapt a foot into the air.

"Got it!" he'd shout, skidding across the boat's wet aluminum floor as I throttled down. Sometimes we'd chase a moth for nearly a minute only to see a passing swallow swoop down out of nowhere to snatch it from the air. Almost all of our captures had little wing nicks. Some even had chunks taken out of their bodies.

With the boat idling, I removed the moth from the net while Robert measured the wind speed and, with a compass, took the moth's directional bearing. The moth looked like a swallowtail butterfly — black with iridescent green bars across each wing — and resembled a flying wedge when its wings were fully open.

Robert wrote his data in a small Rite-in-the-Rain notebook, whose pages sparkled with the moths' black scales. I'd sex the creatures by squeezing their abdomen. Females were fatter, their abdomens filled with lipids — squishy yellow fat. The males had a more pointed rear end. Fatter moths, presumably, had longer journeys ahead of them.

Neil Smith, a STRI scientist, has suggested that *Urania*, which feed on *Omphalea* vines, cue the plant to produce toxic secondary compounds. After several generations of *Urania* attack, the *Omphalea* vines in an area become unpalatable to *Urania*, thus triggering their migration.

Robert thought Smith's idea had a number of problems. "First of all," he said, "most unpalatable insects sequester toxins from their host plants, and herbivory in general induces morphological and chemical responses in plants. But only a very few unpalatable insects migrate. And why would *Urania* migrate unidirectionally, and at the same time every year?" In any event, he concluded, no data are available to test any specific predictions of his ideas.

To date, no one could say definitively why the moths migrated, where they came from, where they went, or why ninety-five percent of those we captured were female. If scientists had these answers, they could focus conservation efforts on the moths' corridors: *Urania* larvae, given the large number of adults that researchers have observed, are probably a major forest herbivore.

Temperate-zone ecologists have lavished attention on the relatively common monarch butterfly, which travels from

2,000 to 3,000 kilometers between its wintering grounds in Mexico and the eastern United States. Its route was worked out in the 1950s, when scientists marked tens of thousands of monarchs. The tagging effort was laborious, and it required a massive public-education campaign. In Panama, getting campesinos who find a tagged *Urania* to telephone a field office would be difficult. The return rate on the monarch project was just .001.

Robert slid our moths into glassine envelopes, alive, and then into his canvas field bag. If the moths didn't succumb to water loss in there, then hypothermia would surely do them in. Until he got around to measuring and weighing them, Robert stored the moths in the freezer.

One day when Robert was too excited to close his field bag, two moths wriggled free of their envelopes and flapped away across the lake. It seemed to happen in slow motion, and I did nothing to halt the exodus. As a symbol of freedom the escape was terribly clichéd, but it made me smile nonetheless. After all these months among scientists, I was still rooting for the entity that defies quantification.

Robert didn't see it that way. He frowned and said, "Now we need two more, in order to complete today's data set."

At week's end, the data set went to pieces. Robert perched on a stool in the instrument lab, which had an herbal, green-leaf smell. First he crushed with his fingers the thorax of this morning's catch, if they still breathed. Working like an automaton, he placed the creatures, one by one, on an electronic balance to get their mass. With tiny scissors he snipped off wings, head, thorax, and legs. He weighed the abdomen, measured the right forewing, and weighed the thorax, which gave him a measure of the insect's muscle mass. In columns, he recorded his numbers in antlike script. Back in Austin a researcher would scan the moth wings to measure their area. Perhaps there was a correlation between wing area and flight speed.

The moth abdomens spent a day in the freeze-dryer, an unimpressive-looking $5,000 box that removes moisture from whatever is put inside. Then Robert attached a vacuum aspirator to a faucet and ran its rubber tubing through a flask and up to a funnel lined with ashless filter paper. In a mortar, he mashed the dried abdomens with a mixture of chloroform and methanol, which yielded a yellow brew of scales and exoskeleton. Robert emptied the mortar into the funnel, and the water vacuum sucked the liquid into the flask. The filter now held pure *Urania* abdominal cuticle, which Robert dried in an oven and then weighed. He subtracted this figure from the original weight of the abdomen and got . . . lipid content. What Robert would do with his hard-won information was unclear to him, at that moment. "But it's cool to know," he said, smiling.

Back in Austin, Robert would make a chart that looked like this:

Year	Sex	Body Mass	Mean Airspeed	Lipid Content
1987	F (*N*=14)	.401 (.265–.599)	3.7 (2.4–4.5)	12.1 (5.3–19.1)
1998	F (*N*=40)	.522 (.255–.839)	3.9 (3.0–4.7)	16.0 (4.0–31.3)
1987	M (*N*=13)	.278 (.222–.328)	4.1 (2.9–5.3)	12.0 (3.7–21.5)
1998	M (*N*=8)	.347 (.314–.407)	3.6 (2.7–4.1)	13.5 (5.8–23.7)
1998	(damaged)			
	F (*N*=17)	.526 (.294–.755)	3.7 (2.7–4.8)	28.1 (18.2–38.3)

The data from this set and data acquired in other locations told Robert that as *Urania* fly farther along their migration route, they tend to slow down: relative lipid content declined simultaneously with flight speeds. Surprisingly, the damaged females from the above set weighed about the same and flew at about the same speed as the undamaged females. Moreover, they had a lipid content almost twice as high. Were predators grabbing the zaftig females, the "calorically more valuable prey"? Robert didn't know for sure, but the results of this little experiment gave him one more question to ponder when he returned to Panama the following year.

Measuring lipid deposition — how much fat a moth carried around — gave researchers an idea of the distance moths flew without eating. While science knew a great deal about the energy balance of migrating birds, there was virtually no information about airspeed and lipids for any migratory insect. In the case of *Urania*, the gap was significant, as the insects migrated in great numbers, consumed nectar along their mysterious route, and made a nice meal for mangrove swallows, which we often saw diving for moths over the lake.

Robert worked with money from the NSF and from the National Geographic Society. In previous years he'd examined butterfly wingbeat frequency on video and quantified butterflies' erratic flight paths. He and colleagues had measured the airspeed of palatable and unpalatable butterflies, then stuck a thermocouple into their thoraxes to measure the temperature of their flight muscles.

It turned out that the thoracic temperature of palatable butterflies was greater than the thoracic temperature of unpalatable butterflies. Higher temperatures meant bigger muscles. In retrospect, it wasn't surprising that butterflies that don't taste bad have bigger flight muscles: they need to do a lot more escaping.

Urania weren't especially quick or unpredictable flyers, which offered a clue to their degree of unpalatability. If predators knew that the moth tasted bad, there was no need for it to hurry or to perfect elaborate evasive maneuvers. *Urania* caterpillars did, in fact, sequester potentially toxic polyhydroxy alkaloids from their host plants. Still, swallows went after the moths with gusto, while jacamars and tyrannid flycatchers ignored them.

The complete *Urania* story was still unwritten, but it promised tantalizing clues to the evolutionary process. What competitive pressures had pushed *Urania* larvae to sequester toxins, and what pressures had pushed some birds, but not others, to overcome them? That mangrove swallows had possibly learned both to neutralize the moth's chemical defenses and to preferentially select moths with higher caloric value was a remarkably sophisticated behavioral response, one that would reap an insectivorous bird a bonanza of lipids — for at least several weeks a year.

One late afternoon, we went out on the lake and counted more than three hundred *Urania* per minute: it was raining moths. We checked some other locations around the canal, just to see what was happening, and got pelted with moths as we drove. The sun was shining off to the west, illuminating a rainbow in the slate-blue sky. All the moths were headed north. We caught five — boom, boom, boom, boom, boom. All females. Then we noticed something new: the moths were sharing airspace with enormous dragonflies, Bell helicopters to the moths' dainty Cessnas.

"This is crazy!" Robert shouted above the drone of the motor. "This is just crazy!" He was delighted — with the number of moths and with the dragonflies, which were flying far too fast to capture.

Robert had never seen moths migrating along with dragonflies. But the observation gave him a righteous intellectual thrill because there was an idea popular among some tropical biologists that migrations were set in motion by the flowering of certain trees, which set up seasonal rhythms that dictated the behavior of organisms at all trophic levels. Robert hadn't bought it. "Oh, come on!" he said. "I don't think so. These are carnivorous insects." Moreover, why would female moths, vulnerable in their transit, subject themselves to the mandibles of one of the fastest-flying and most predacious insects in the world? Give a dragonfly thirty minutes and it could eat its own weight in food. The insect's bulging eyes occupied most of its

head, which may have been why we had such rotten luck netting even one.

The sun began to set, and Robert laid his net down in the
boat. "Let's go back to the lab," he said, taking a final look
around. "I don't like being out here with all these carnivores."

In a contemplative mood as we zipped home, I wrapped my
arms over my belly. Would my daughter ever see the *Urania*
migration? Before I became pregnant, the idea of conserving
biodiversity for the children of the future had been just a plati-
tude to me, a noble goal. But it didn't compel me. Now, sitting
on the lake amid thousands of brilliant moths, the idea that
these creatures might not be around in another generation
seemed like a real, and very sad, possibility.

Though I had at first resisted her, my little chimp was finally
growing on me — much as the spider monkeys had grown on
Chrissy, once she'd learned to recognize them. And because I
now had feelings for my girl, I wanted her to enjoy the natural
world the way that I had.

Back north, scientists were spending hundreds of millions of
dollars parsing the human genome and building cellular structures. But in what kind of world, I wondered, would we make
use of such information and innovations? Would there be wild
lands and clean lakes for my daughter to enjoy? Would there
be migrations of monarchs? The bugling of elk, the howling of
wolves? While we spend billions on military security every
year, I reflected, we spend a mere pittance on what I considered a human right: a protected environment.

For the first time in history, the fate of the planet — what it
would look like, whom it would shelter, how it would smell —
lay entirely in our hands. Human beings, and not a meteor or
glaciers, would decide. Prognosticators said that the most profound challenge of the new century would be maintaining a
livable earth. But "livable" wasn't enough for me. I was greedy.
For my daughter I wanted open space, I wanted fast-flowing

waterways and forested mountains. I wanted her to witness, if she so desired, an iridescent green and black river of migrating diurnal moths.

It was the middle of October now, normally the wettest of months, and the weather was strangely dry. Almost all the usual humidity was absent, and we were enjoying an Indian summer of warm air and blue skies.

It didn't take long to figure out what was going on. Our delightful days came at the expense of several million people to the west and north. A low-pressure system was working its way through the Caribbean, sucking Panamanian humidity into the maw of a disturbance that would culminate in Hurricane Mitch, the most destructive storm of the twentieth century. A Class Five hurricane, Mitch would eventually kill more than eleven thousand people in Honduras, Nicaragua, El Salvador, and Guatemala.

My husband, Peter, had flown down to visit me, and in our news vacuum, Panama had never seemed so pleasant. For months, actually, I'd been quite happy here. I had become accustomed to lounging on the raft in the afternoon, chatting with friends while parakeets flocked overhead. I spent hours on the lake, counting butterflies and moths. I watched the monkeys in the trees.

Plotted on a graph, with time on the x axis and contentment on the y, the line that defined my mood would rise at a forty-five-degree angle. My R^2 would be close to 1, with outlying points for the day I received 131 chigger bites, the night the termites streamed over my head while I slept, the times I became disoriented in the forest, and the half-dozen Bambis at which mathematical equations brought tears of frustration to my eyes.

As my departure approached, I could see why E. O. Wilson had said he'd retire on ten thousand dollars a year to live on BCI. Why Robert Dudley longed to build his tropical air castle at the end of the dormitory path. Why Robin Foster had pined for the job of island naturalist. Why so many long-term residents had mixed feelings about leaving.

But whatever the charms of BCI, I had no pretensions about wanting to stay much longer. Perhaps the nesting instinct was finally kicking in. I thought about buying something for my baby in Panama City, but I couldn't imagine what. Instead, I preserved some *Passiflora vitifolia* blossoms for her, pressed the wings of a butterfly into a book, and boiled clean a brocket skull I'd found near Zetek 6. I hoped she would like them.

Constraints were closing in on me: my scientists were leaving the island and my due date approached. As much as I looked forward to being home, though, I yearned to connect just a few of the tiny puzzle pieces of basic research that I'd had the chance to glimpse. But inside me, a different sort of puzzle piece was growing. My new job was to figure out how her demanding pieces could possibly fit into mine.

By now, Bret and his Frisbee were gone, as were Guillermo, Stefan, Sabine, Paul, Greg, Doug, and the Roberts — Stallard and Dudley. Hubi was in the long process of packing. Chrissy had a month and a half left, but she wasn't looking forward to California. There, she had no home, no job, no social life, just month after month of analyzing fecal samples. At home there would be rent and electric bills to pay, meals to cook, parking spots to fight over. The birds, up north, looked dull, the forests depauperate. In comparison, life on BCI was luxurious, and rich.

For a lot of people, BCI represented perpetual summer, a break from the responsibilities of the real world. The work here was good, the forest as compelling a character in their lives as any significant other back home. To live here was to live in a seething cauldron of growing, grasping, competing, cooperating, mating, and dying organisms. Nothing was static, all was change. The forest was mysterious and intricate, flamboyant and dull, weird and gaudy, engaging and indifferent.

For those who returned to the island year after year, studying these contradictions offered endless pleasure and intellectual reward.

It wasn't a novel thought, but the more I learned about the tropical forest, the more I came to appreciate it and the more I wanted to know. By the time I left BCI, my knowledge of its forest was neither integrated nor deep, but that didn't bother me. I took great satisfaction in the wide-ranging natural history I had learned here. I had started on the sort of education that today's narrowly focused scientists have no time to pursue.

Even as I packed my belongings, new residents arrived, others left, and the work of BCI rolled endlessly on. There were old questions to answer: How does evolution work? Why do we see diversity gradients? What is the importance of design-constraint tradeoffs?

And there were new questions as well. The young field of ecology was probing the physiology of plants, the movement of nutrients and water through a forest. Faced with changing global conditions, conservationists needed to know how quickly a disturbed rain forest could recover and what role the tropics played in the world's climate and nutrient cycles. An entire career could be fashioned from studying greenhouse gases.

During my final week I reflected on the past year. Some things had gone wrong for residents, and some things had gone right. Such was the nature of science, whether one toiled in a laboratory or in the tropical rain forest. Progress — if it came at all — wasn't always linear, and researchers often took one step backward for every two that brought them forward.

For established scientists, who had the luxury of returning to BCI year after year, a misstep was but a blip. Slowly but surely they would accumulate their puzzle pieces of data until the day they wiggled them into place (or ran out of money).

For scientists just starting out, the unexpected often provided new opportunities. Hubi, for example, didn't meet all his original goals, but he did come up with solid ant-herbivory numbers, and he helped make an unexpected discovery about a cause of colony movement. Both Chrissy and Bret had problems collecting data. Chrissy would devise a successful dissertation nonetheless; Bret would switch gears and move into the theoretical realm.

Trying to make sense of my own experience on the island, I found myself shifting between the idea that the forest was immeasurably complex and unknowable, and the idea that the forest was intimately measurable and potentially explainable — a dichotomy of surrender and hope. In the end, hope won out: I trusted that the rain forest's mysteries were not occult, and that the tools and the imaginations of those who lived and breathed Barro Colorado Island would eventually — maybe hundreds of years from now — understand the whole glorious mess, the green hell, the riot. And our sense of wonder, in lockstep, would only grow and grow.

Peter stayed with me on BCI for a month. We walked the trails, we caught and released butterflies, and we read Jack Kerouac aloud to our growing fish. Her head pointed one way, then another. She seemed to revel in taking her exercise at night. Maybe she'd grow up to study moths, I thought.

Our final night on the island, Peter and I bought a case of beer and settled into the bar with Chrissy and Hubi and a dozen other residents. Jenny Apple had baked me a cake in the shape of a book. All the Germans — they were now a mob of eight — were smoking. We took a lot of photos, and people kept asking me when I'd be back. They had the idea, probably from watching monkeys in the treetops, that toting a baby around the jungle would be easy.

Feeling a bit overwhelmed and tired, I said good night to the group at eleven. I regretted not having seen Bert, but he surprised me by coming down to the *Jacana* at six the next morning to bid me, in his own undemonstrative manner, a fond farewell. "Be of good cheer," he said. I would miss him in the months to come.

Back in New York, I kept in touch with island residents through e-mail. I learned that the wind had blown over one of the floating docks, taking a number of boats with it. STRI was collaborating with hoteliers to build an ecotourism resort in Gamboa. Ira Rubinoff was convinced that teaching tourists about Panama's biodiversity, with the help of scientist guides, was the key to its preservation. Two harpy eagles were released on BCI: top predators with two-meter wingspans, they immediately began picking sloths and howler monkeys out of the trees. Katie Milton, I heard, was not pleased.

Slowly the stream of e-mail thinned, and then there was nothing. All my favorite island residents had returned to their universities — to sort insects, crunch numbers, and defend their dissertations. Their stories, for me, had ended. I had to make my own peace with the idea that my final connections to the island were gone.

At home, the noise of the city traffic bothered me at first, and the air smelled especially dirty. Gradually, I got used to it, and soon it was as if I'd never left. I was happy to be home, though the days were now short and the sky leaden. I was starting anew. A baby grew in my belly — my own evolutionary product — and I felt, as I never had before, part of the great continuum of life.

Epilogue

ANYTHING COULD HAPPEN at the end of a field season. One day Chrissy was in the forest, scraping monkey dung off leaves. The next day she was done, and over it. She had six days until her flight home, but she couldn't stand to look at another monkey. Residents threw her not one but two going-away parties.

At Berkeley, Chrissy's much-anticipated mental collapse, foretold on a rock near Zetek 5, failed to materialize. She married her longtime boyfriend, analyzed her samples, and wrote her dissertation. By September, Dr. Campbell was teaching at Santa Monica College.

Hubi left BCI in October. Back in Germany, he began to write up his work and teach experimental plant ecophysiology to graduate students at his university. The siren song of the ants still called, and he sought a postdoctoral fellowship that would let him explore the trophic interactions between leaf cutters, plants, fungi, predators, and parasites and investigate what role biodiversity played in these interactions.

At the end of the rainy season, Sabine went home to identify and count her arthropods. She returned to Panama for another dry-season sweep of her traps, then started her thesis. She was interrupted by a job offer: as soon as she got her degree, Germany's Channel ZDF, in Munich, wanted her to produce and edit natural-history programs.

After a frenzy of stuffing laundry into bags and packing up equipment, Bret left BCI in the middle of August. He married Heather Heying a week later and worked on his dissertation in Madagascar, where she was studying frogs. Back in Michigan, the Museum of Zoology presented him with an award for high achievement in ecology and evolution.

Although *Nature* rejected Bret's telomere paper, Lissy Coley, at the University of Utah, offered Bret a postdoc based on his novel ideas about tradeoffs and their application to biomedical research. She wasn't sure what work Bret would do in Utah, but she believed in his intelligence and creativity, and she didn't want him to fall through the cracks. She believed that Bret's theoretical work was going to bring him a lot of attention, and that he would, someday, change the way people think.

Bibliography

These are some of the books I read while researching *The Tapir's Morning Bath*. Some relate to Barro Colorado Island and Panama; others are more generally about tropical rain forests and biology.

Allee, Warder C., and Marjorie Hill Allee. *Jungle Island*. Chicago: Rand McNally, 1925.

Angehr, George R. *Parting the Green Curtain: The Evolution of Tropical Biology in Panama*. Washington, D.C.: Smithsonian Institution, 1989.

Attenborough, David. *Life on Earth*. Boston: Little, Brown, 1979.

Barbour, Thomas. *Naturalist at Large*. Boston: Little, Brown, 1943.

Bates, Henry Walter. *The Naturalist on the River Amazons*. 1863; reprint: New York: Penguin Books, 1989.

Carpenter, C. R. *A Field Study of the Behavior and Social Relations of Howling Monkeys*. Baltimore: Johns Hopkins University Press, 1934.

Chapman, Frank M. *My Tropical Air Castle*. New York: D. Appleton, 1929.

———. *Life in an Air Castle*. New York: D. Appleton, 1938.

Darwin, Charles. *The Voyage of the Beagle*. 1845; reprint: New York: Bantam, 1958.

Dawkins, Richard. *The Blind Watchmaker*. New York: Norton, 1987.

de Waal, Franz. *Chimpanzee Politics.* Baltimore: Johns Hopkins University Press, 1989.

Douglas, Marjory Stoneman. *Adventures in a Green World: The Story of David Fairchild and Barbour Lathrop.* Miami: Field Research Projects, 1973.

Emmons, Louise H. *Neotropical Rainforest Mammals: A Field Guide.* Chicago: University of Chicago Press, 1990.

Evans, Howard Ensign and Mary Alice Evans. *William Morton Wheeler, Biologist.* Cambridge, Mass.: Harvard University Press, 1970.

Fairchild, David. "The Jungles of Panama." *National Geographic* 41 (1922).

———. "BCI Laboratory." *Journal of Heredity*, March 1924.

———. *The World Was My Garden: Travels of a Plant Explorer.* New York: Charles Scribner's Sons, 1938.

Forsyth, Adrian, and Ken Miyata. *Tropical Nature: Life and Death in the Rain Forests of Central and South America.* New York: Touchstone, 1984.

Fortey, Richard. *Life: A Natural History of the First Four Billion Years of Life on Earth.* New York: Knopf, 1998.

Futuyma, Douglas J. "Wherefore and Whither the Naturalist?" *American Naturalist*, January 1998.

Goodall, Jane. *In the Shadow of Man.* Boston: Houghton Mifflin, 1988.

Grant, Susan. *Beauty and the Beast: The Coevolution of Plants and Animals.* New York: Charles Scribner's Sons, 1984.

Grice, Gordon. *The Red Hourglass: Lives of the Predators.* New York: Delacorte Press, 1998.

Gross, Alfred O. "BCI Biological Station." *Smithsonian Report*, 1926, pp. 327–42.

Hölldobler, Bert, and Edward O. Wilson. *The Ants.* Cambridge, Mass.: Harvard University Press, 1990.

Jolly, Allison. *Lucy's Legacy: Sex and Intelligence in Evolution.* Cambridge, Mass.: Harvard University Press, 1999.

Kricher, John C. *A Neotropical Companion.* Princeton: Princeton University Press, 1989.

Leigh, Egbert G., Jr., Stanley Rand, and Donald M. Windsor, eds. *The Ecology of a Tropical Forest: Seasonal Rhythms and Long-term*

Changes. Washington, D.C.: Smithsonian Institution, 1982; 2nd ed., 1996.

Leigh, Egbert Giles, Jr. *Tropical Forest Ecology: A View from Barro Colorado Island.* New York: Oxford University Press, 1999.

Lewin, Roger. *Thread of Life: The Smithsonian Looks at Evolution.* Washington, D.C.: Smithsonian Books, 1982.

Lowman, Margaret D. *Life in the Treetops: Adventures of a Woman in Field Biology.* New Haven: Yale University Press, 1999.

McCullough, David. *The Path Between the Seas: The Creation of the Panama Canal 1870–1914.* New York: Simon & Schuster, 1977.

Moffett, Mark W. *The High Frontier: Exploring the Tropical Rainforest Canopy.* Cambridge, Mass.: Harvard University Press, 1993.

Raby, Peter. *Bright Paradise: Victorian Scientific Travellers.* Princeton: Princeton University Press, 1996.

Ridley, Mark, ed. *The Darwin Reader.* New York: Norton, 1987.

Sapolsky, Robert M. *The Trouble with Testosterone: And Other Essays on the Biology of the Human Predicament.* New York: Scribner, 1997.

Vermeij, Geerat. *Privileged Hands: A Remarkable Scientific Life.* New York: W. H. Freeman, 1997.

Wallace, Alfred Russel. *The Malay Archipelago.* 1869. Reprint: New York: Oxford University Press, 1987.

Wallace, David Rains. *The Monkey's Bridge: Mysteries of Evolution in Central America.* San Francisco: Sierra Club Books, 1997.

Weiner, Jonathan. *The Beak of the Finch.* New York: Vintage, 1995.

———. *Time, Love, Memory: A Great Biologist and His Quest for the Origins of Behavior.* New York: Vintage, 1999.

Wilson, Edward O. *Naturalist.* New York: Warner Books, 1995.

Wong, Marina, and Jorge Ventocilla. *A Day on Barro Colorado Island.* Panama: Smithsonian Tropical Research Institute, 1995.

In addition to the above books, I found the papers concerning BCI, stored in boxes RU 134 and RU 135 at the Smithsonian Institution Archives in Washington, D.C., extremely useful, as were the Smithsonian's oral history interviews with Ira Rubinoff, Stan Rand, Nicholas Smythe, Don Windsor, Fausto Bocanegra, Graham Bell Fairchild, Oscar Dean Kidd, and Alexander Wetmore.

Acknowledgments

This book wouldn't have been possible without the generosity and kindness of Barro Colorado's resident scientists, in particular those who opened their lives and their labs to me on a daily basis. Those who tolerated my constant curiosity include Christina Campbell, Robert Dudley, Hubi Herz, Sabine Stuntz, and Bret Weinstein. I subjected Greg Adler, Rafael Aizprua, Lissy Coley, Diane DeStevens, Robin Foster, Guillermo Goldstein, Nelida Gomez, Gerald Heckel, Heather Heying, Kristina Huffington, Jennifer Johns, Elisabeth Kalko, David Marsh, Kiersten Montague, Doug Schemske, Stefan Schnitzer, Neil Smith, Bob Srygley, Robert Stallard, Paul Trebe, Joe Wright, and Jayne Yack to somewhat less-intense interrogation.

In addition to those listed above, I'd like to thank Ira Rubinoff, director of the Smithsonian Tropical Research Institute, for welcoming a writer to live among his flock; Oris Acevedo and Daniel Milan, for helping me get situated on the island; and Egbert Leigh, for sharing his counsel, his passion for the forest, and his bourbon.

This book is about biologists and their work. It contains a great many facts, and I tried to be faithful to what researchers told me about their projects and showed me in the field. Some have had the opportunity to review what I've written about them; for those who haven't, I hope sincerely that I've done their investigations justice. Those who run experiments in the wild, my sources tell me, are content with a ten percent data variance. I aimed higher, but I accept full responsibility for any errors that slipped through the cracks.

*

For encouraging me to write this book I'm grateful to my literary agent, Heather Schroder of ICM, and to Robert Dudley. I'd like to thank Pamela Henson, at the Smithsonian Institution Archives, for her help in obtaining research materials. For their moral and financial support, I thank the Alicia Patterson Foundation, which named me a Fellow in 1999. For inspiring me to revisit BCI — and for saving me, long ago, from the clutches of a giant *Blaberus* cockroach — I thank Edward O. Wilson. For bringing me into the woods and opening my eyes to nature, I'd like to acknowledge my brother, Joshua Royte — my very first naturalist.

I owe much to my editor, Anton Mueller, who understood what I was writing about long before I was convinced. I offer snappy salutes to those who commented on my manuscript as it evolved — in particular, to the insightful Lisa Chase, the perspicacious John Seabrook, and the indefatigable Bret Weinstein. Katya Rice copyedited the book with intelligence and wit; I'm lucky to have had her looking over my shoulder. Most especially, I thank my husband, Peter Kreutzer, who read, commented, and listened with extraordinary patience throughout the book's gestation.

Last but not least, I thank Lucy, who napped.